Byproducts, Waste Biomass and Products to form Green Diesel and Biocrude Oils

Byproducts, Waste Biomass and Products to form Green Diesel and Biocrude Oils

Editors

Brajendra K. Sharma
Kirtika Kohli

MDPI • Basel • Beijing • Wuhan • Barcelona • Belgrade • Manchester • Tokyo • Cluj • Tianjin

Editors
Brajendra K. Sharma
Prairie Research Institute -
Illinois Sustainable Technology
Center, University of Illinois at
Urbana-Champaign
USA

Kirtika Kohli
Prairie Research Institute -
Illinois Sustainable Technology
Center, University of Illinois at
Urbana-Champaign
USA

Editorial Office
MDPI
St. Alban-Anlage 66
4052 Basel, Switzerland

This is a reprint of articles from the Special Issue published online in the open access journal *Energies* (ISSN 1996-1073) (available at: https://www.mdpi.com/journal/energies/special_issues/BWBP_GDBCO).

For citation purposes, cite each article independently as indicated on the article page online and as indicated below:

LastName, A.A.; LastName, B.B.; LastName, C.C. Article Title. *Journal Name* **Year**, *Article Number*, Page Range.

ISBN 978-3-03943-517-3 (Hbk)
ISBN 978-3-03943-518-0 (PDF)

© 2020 by the authors. Articles in this book are Open Access and distributed under the Creative Commons Attribution (CC BY) license, which allows users to download, copy and build upon published articles, as long as the author and publisher are properly credited, which ensures maximum dissemination and a wider impact of our publications.

The book as a whole is distributed by MDPI under the terms and conditions of the Creative Commons license CC BY-NC-ND.

Contents

About the Editors vii

Preface to "Byproducts, Waste Biomass and Products to form Green Diesel and Biocrude Oils" ix

Inam Ullah Khan, Zhenhua Yan and Jun Chen
Production and Characterization of Biodiesel Derived from a Novel Source *Koelreuteria paniculata* Seed Oil
Reprinted from: *Energies* **2020**, *13*, 791, doi:10.3390/en13040791 1

Thao Nguyen Luu, Zouheir Alsafra, Amélie Corato, Daniele Corsaro, Hung Anh Le, Gauthier Eppe and Claire Remacle
Isolation and Characterization of Two Microalgal Isolates from Vietnam with Potential for Food, Feed, and Biodiesel Production
Reprinted from: *Energies* **2020**, *13*, 898, doi:10.3390/en13040898 17

Attada Yerrayya, A. K. Shree Vishnu, S. Shreyas, S. R. Chakravarthy and Ravikrishnan Vinu
Hydrothermal Liquefaction of Rice Straw Using Methanol as Co-Solvent
Reprinted from: *Energies* **2020**, *13*, 2618, doi:10.3390/en13102618 33

Szymon Szufa, Grzegorz Wielgosiński, Piotr Piersa, Justyna Czerwińska, Maria Dzikuć, Łukasz Adrian, Wiktoria Lewandowska and Marta Marczak
Torrefaction of Straw from Oats and Maize for Use as a Fuel and Additive to Organic Fertilizers—TGA Analysis, Kinetics as Products for Agricultural Purposes
Reprinted from: *Energies* **2020**, *13*, 2064, doi:10.3390/en13082064 53

Changzhou Chen, Peng Liu, Minghao Zhou, Brajendra K. Sharma and Jianchun Jiang
Selective Hydrogenation of Phenol to Cyclohexanol over Ni/CNT in the Absence of External Hydrogen
Reprinted from: *Energies* **2020**, *13*, 846, doi:10.3390/en13040846 83

Nikolaos Montesantos and Marco Maschietti
Supercritical Carbon Dioxide Extraction of Lignocellulosic Bio-Oils: The Potential of Fuel Upgrading and Chemical Recovery
Reprinted from: *Energies* **2020**, *13*, 1600, doi:10.3390/en13071600 95

Laura Brenes-Peralta, María F. Jiménez-Morales, Rooel Campos-Rodríguez, Fabio De Menna and Matteo Vittuari
Decision-Making Process in the Circular Economy: A Case Study on University Food Waste-to-Energy Actions in Latin America
Reprinted from: *Energies* **2020**, *13*, 2291, doi:10.3390/en13092291 131

About the Editors

Brajendra K. Sharma is a Sr. Research Scientist at the Illinois Sustainable Technology Center, a division of the Prairie Research Institute, University of Illinois at Urbana-Champaign (UIUC). He is also a faculty member of the Dept. of Agricultural and Biological Engineering at UIUC and an Adjunct Professor of Energy and Environmental Systems at NCAT State University, North Carolina. He is a fellow of the Royal Society of Chemistry and the Society of Tribology & Lubrication Engineers. He is on the Editorial Board of *Fuel Processing Technology* and is the associate technical editor for the *Journal of Surfactants and Detergents*. His areas of interest are biobased materials like biolubricants, biobased asphalt binders, biobased additives and chemicals, biofuels, and alternative fuels.

Kirtika Kohli is a Post-Doctoral Researcher at the Illinois Sustainable Technology Center, a division of the Prairie Research Institute, University of Illinois at Urbana-Champaign (UIUC). She earned her Ph.D. in Chemical Sciences from the CSIR–Indian Institute of Petroleum, Dehradun, India (title of thesis *Deactivation Study of Residue Hydroprocessing Catalysts*). Her research interests are multidisciplinary, and include the synthesis and detailed characterization of homogeneous and heterogeneous catalysts, catalytic deactivation mechanisms, hydrotreating and the hydrocracking of petroleum feedstocks and biocrude oils, lignin extraction from waste biomass feedstocks, development of bio-energy and high-value bioproducts from bioresources via thermochemical and catalytic conversion processes, CO_2 valorization to chemicals, and plastic waste conversion processes to produce fuels and chemicals. She is the author of more than 15 papers in catalyst synthesis for heavy crude oil processing, lignin extraction from waste biomass feeds, and plastic waste conversion to fuels. She had two granted US patents for developing novel dispersed catalysts for slurry phase hydrocracking and technology for using plastic waste as a hydrogen donor for heavy crude oils and residues.

Preface to "Byproducts, Waste Biomass and Products to form Green Diesel and Biocrude Oils"

The increasing interest in clean technologies, renewable energy resources and their products to reduce the dependence on fossil fuels and reducing the effect of greenhouse gas emissions is gaining momentum. Renewable diesel and biocrude oils are promising as an alternative to fossil oils. These products will have a significant share in future global energy portfolios and in reducing greenhouse gas emissions. The conversion technologies, system integration approaches, and life cycle impacts of bio-derived fuels can vary widely because of the large diversity of biomass feedstocks. However, the main challenges associated with biomass conversion processes are feedstock variability, the pre-treatment processes involved, land use concerns, and high production costs, which hinder their broad-scale market acceptance. Thus, new conversion technologies are expected to increase production potential by allowing for the use of an array of waste biomass feedstocks (agricultural residues, forest residues, and industrial residues) and byproducts produced from current biomass conversion processes.

The purpose of the Special Issue "Byproduct, Waste Biomass and Products to form Green Diesel and Biocrude Oils" of the journal *Energies* was to provide a collection of papers that reflected modern topics of research and new developments in the field of renewable diesel and biocrude oils from waste biomass resources. This book includes seven peer-reviewed articles covering a review article "supercritical CO_2 extraction of bio-oils" and original research articles. The topics include the hydrothermal liquefaction of rice straw, the torrefaction of oats and maize straws; the utilization of microalgal isolates, selective hydrogenation, and novel non-edible plant seed oil from *Koelreuteria paniculate* to produce biodiesel and chemicals. Thus, the collection of research articles provides comprehensive details on the current and emerging feedstocks, and advanced conversion technologies for the development of sustainable energy resources. In addition, one research article also describes the decision-making process involved in the circular economy of food-waste to energy actions in Latin America. This volume will be of use as a reference to chemists and engineers in both academia and industry that would like to develop an understanding of biomass conversion processes, biodiesel production, and energy recovery from waste biomass feedstocks, as well as to students.

We thank the journal *Energies* for the opportunity to publish this book.

Brajendra K. Sharma, Kirtika Kohli
Editors

Article

Production and Characterization of Biodiesel Derived from a Novel Source *Koelreuteria paniculata* Seed Oil

Inam Ullah Khan, Zhenhua Yan * and Jun Chen *

Key Laboratory of Advanced Energy Materials Chemistry (Ministry of Education), Renewable Energy Conversion and Storage Center, College of Chemistry, Nankai University, Tianjin 300071, China; 1120176003@mail.nankai.edu.cn
* Correspondence: yzh@nankai.edu.cn (Z.Y.); chenabc@nankai.edu.cn (J.C.)

Received: 2 January 2020; Accepted: 5 February 2020; Published: 11 February 2020

Abstract: Biodiesel is a clean and renewable fuel, which is considered as the best alternative to diesel fuel, but the feedstock contributes more than 70% of the cost. The most important constituent essential for biodiesel development is to explore cheap feedstock with high oil content. In this work, we found novel non-edible plant seeds of *Koelreuteria paniculata* (KP) with high oil contents of 28–30 wt.% and low free fatty acid contents (0.91%), which can serve as a promising feedstock for biodiesel production. KP seed oil can convert into biodiesel/fatty acid methyl esters (FAMEs) by base-catalyzed transesterification with the highest biodiesel production of 95.2% after an optimization process. We obtained the optimal transesterification conditions, i.e., oil/methanol ratio (6:1), catalyst concentration (0.32), reaction temperature (65 °C), stirring rate (700 rpm), and reaction time (80 min). The physico-chemical properties and composition of the FAME were investigated and compared with mineral diesel. The synthesized esters were confirmed and characterized by the application of NMR (^{1}H and ^{13}C), FTIR, and GC-MS. The biofuel produced from KP seed oil satisfies the conditions verbalized by ASTM D6751 and EN14214 standards. Accordingly, KP source oil can be presented as a novel raw material for biofuel fabrication.

Keywords: *Koelreuteria paniculata* biodiesel; non-edible feedstock; transesterification; physicochemical characterization; optimization

1. Introduction

With the rising demand for energy, attention has been focused on alternate renewable fuel resources besides fossil fuels [1]. The research impetus of renewable fuel resources is on reduced net CO_2 emission, thus controlling air, soil, and water pollution, and ultimately minimizing the well-being threat [2–4]. Biomass is the plant material derived from the reaction between CO_2 in the air, water, and sunlight, via photosynthesis to produce carbohydrates that convert solar energy to chemical energy. About 170 billion tons of biomass produces every year on earth [5]. Biomass has always been a major source of energy for mankind; more than 90% of the living energy of about 2.5 billion people in the world is biomass energy. Biomass energy has the advantages of easy combustion, less pollution, and lower ash content, it has been the main source of human energy. The currently accounts for an estimated 10%–14% of the world's energy supply and more than 60% in underdeveloped regions, more than 90% of the living energy of about 2.5 billion people in the world is biomass energy [6–8]. However, the biomass energies are low calorific value and thermal efficiency, large volume, not easy to transport, and the thermal efficiency of direct combustion biomass is as low as 10% to 15%. To increase the utilization of energy and make it more valuable, biomass energy can be converted into liquid and gaseous fuels as biodiesel through chemical routes [9–11]. The biodiesel fuels are oil esters [9] which have been an auspicious substitute source to fossil fuels because it is biodegradability, renewable nature, low emission lethal and good storing and transportation properties [2,12]. In recent years, the exploration and development of valuable energy plant seeds as high-efficiency biodiesel sources have become a research hotspot [13].

Currently, more than 95% of biodiesel is produced from edible oils, which seriously comprising deforestation and damage to wildlife [14] and reducing essential oil resources for the growing human population. Even though edible oils give enormous biodiesel production and easily transesterified because of their less free fatty acid contents [12,15,16]. However, they are reducing essential oil resources for the growing human population in many countries and regions; the use of edible oils for biodiesel production will direct competition with food uses [17]. The high cost of the raw material of biodiesel is the major barrier for its commercialization [18]. To control this situation, it is important to explore non-edible oil resources, which can be proved as cost-effective as well as resolving the food concerns [19]. In addition, the production of non-edible oils does not require ideal conditions [13]. Even non-fertile land, uncultivated land, road/field boundaries, disgraced forests, and irrigation canals can be used to produce non-edible oil crops [20]. Considering cost effectiveness, non-edible oil raw materials have been used in the production of biodiesel in various studies, including *Croton megalocarpus* [21], *Prunus dulcis* [22], *Prunus sibirica* [18], *Rhazya stricta* Decne [23], rubber seed oil [24], *Silybum marianum* L. [25], wild *Brassica Juncea* L. [26], etc. However, the challenge remains to produce high-quality biodiesel from cheap and available non-edible sources [27].

Koelreuteria paniculata (KP) belongs to the family Sapindaceae is a novel non-edible seed oil source that can be explored for biodiesel production. KP is rarely attacked by pests and grows in a wide range of soils, including high pH soils. The inflorescences are very remarkable and contain ellipsoid pods, which can produce huge amounts of seeds. A single average plant produces 15–20 kg seeds annually. Four to five hundred of KP trees can be planted per hectare area. Approximately one-hectare area will produce 115,000 kg seeds, the productivity of oil is about 30,000 kg per hectare area. KP species are originally from China, Japan, and Korea. It's been used as an ornamental plant and has no further uses. The KP tree is considered invasive in several areas due to the prolific seed production and bazillions of offspring. Due to all such characteristics, it is best suits to grown in waste and barren lands and use its seeds for alternative energy, which have 28%–30% oil contents (detailed descriptions are present in Supplementary Materials). However, to the best of our knowledge, the investigating KP seeds is a potential novel source of biodiesel.

In this work, we explored the efficacy of KP for the production and assessment of seed oil as biodiesels through a small budget and easy technique base-catalyzed transesterification method. The important parameters and their effects on the production and the fatty acid methyl ester (FAMEs) contents were optimized, comprising the oil to methanol molar ratio, KOH conc., temperature, time, and stirring intensity. The ester conversion and fatty acid composition were confirmed by Fourier transform infrared spectroscopy (FTIR), nuclear magnetic resonance (NMR) and gas chromatography-mass spectrometry (GC-MS). Its physical and chemical properties (density, kinematic viscosity, cloud point, cloud point, flash point, ignition point, etc.) were studied and compared with ASTM D6751 and EN14214. ICP-OES (inductively coupled plasma atomic emission spectrometry) and elements analyzer (EA) were used for elemental analysis, and the results showed that biodiesel was environmentally friendly.

2. Materials and Methods

2.1. Materials

Koelreuteria paniculata seeds used in the current research work were collected from the wild field in Binhai new area near Nankai University's new campus Tianjin, China, during months in November 2018. To remove the foreign matter, such as dirt, dust, chaff, and stones, as well as broken and immature seeds; the seeds were clean manually. As per instruction to control its original form and quality before any conditioning sealed in plastic bags at an ambient temperature of 25 ± 3 °C.

2.2. Chemicals

Methanol 99–100%, ethanol 99.8%, ethanol 99%–100%, hydrochloric acid 37%, *n*-hexane 96% (Sinopharm Chemical Reagent Co., Ltd., Shanghai, China), cyclohexane (Macklin), ≥99.9%, chloroform-d (Fluorochem, Hadfield, UK), acetonitrile, petroleum ether, phenolphthalein pH 8.2–9.8 (Sigma Aldrich, St. Louis, MO, USA), sodium hydroxide pellets ≥98%, potassium hydroxide pellets 99.99%, starch, acetone (99%), sodium thiosulphate (99.0%), sulphuric acid 98%, (Aladdin, Shanghai, China), isopropanol, and bromine were obtained from Merck, Darmstadt, Germany.

2.3. Preparation of Feedstock's/Soxhlet Extraction

Solvent extraction is the process in which the oil is removed from seed through a liquid solvent, known as leaching. The KP were dried at 45 °C for 22 h after washing with deionized water to remove the dust (to completely finish the moisture), then ground with electrical grinder machine (XIANTAOPAI XTP-10000A, China). The oil was extracted by petroleum ether in a Soxhlet apparatus for 8 h [28], and the solvent was distilled under vacuum by a rotary evaporator (TOKYO RIKAKIKAI Co. Ltd., Tokyo, Japan, N-1210B). To remove the remaining moisture the obtained oil was dried with anhydrous sodium sulfate; then, for supplementary calculation and consumption, the obtained oil was filtered and put in a wrapped container at 5 °C.

2.4. Experimental Procedure of Transesterification

The transesterification (Figure 1) was done in a round bottom 2 L flask furnished with a magnetic stirrer, reflux condenser, sampling outlet, and thermometer. Before started the reaction, 500 mL seed oil was hated at 70 °C. The reaction was done with a catalyst conc. (KOH) 0.32% w/w of the oil and oil to methanol molar ratio was 6:1. The freshly prepared 83.33 mL potassium methoxide solution was added in oil at 70 °C temperature and a constant stirring speed of 700 rpm [29]. To confirm the comprehensive transformation of TAG to FAMEs, the stirring was continued for 80 min. The mixture was relocated in a separatory tube funnel to separate the biodiesel and glycerin layers. The dense glycerin layer was detached. The upper biodiesel layer was then washed two to three times with hot distilled water to make it clean and clear. The product was dried through anhydrous sodium sulfate and evaporated the remaining methanol at 50 °C.

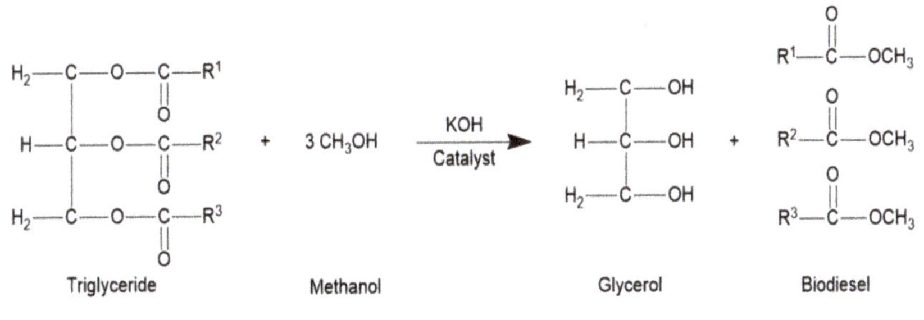

Figure 1. Transesterification of *Koelreuteria paniculata* oil to biodiesel.

The product yield was premeditated by Equation (1),

$$\text{Yield\%} = \frac{\text{Gram of biodiesel produced}}{\text{Gram of oil used}} \times 100 \qquad (1)$$

2.5. Analytical Study of KP Biodiesel

The KP biodiesel (KPBD) produced through transesterification was examined for its eventual study using an EA (Vario EL CUBE, Hanau, Germany), which delivers C, H, and N outcomes, whereas the O value was examined from the difference. When the summation of C, H, and N values is deducted from

100, it gives the O proportion, and for complete element, a study has been used ICP-OES (Spectro-blue, Kleve, Germany). The procedure is presented in Supplementary Materials Figure S3. The produced biodiesel was checked and verified with the Fourier transformed infrared spectrophotometer (Bruker vertex 70 FT-IR Spectrometer, Karlsruhe, Germany) with a resolution of 1 cm^{-1} and 15 scans, in the spectra range of 400–4000 cm^{-1}. To describe various functional groups that originate in the found biodiesel, Nujol mull was used as a reference. NMR spectra were obtained with CDCl$_3$ as a solvent on a (Bruker Avance III400 NMR Spectrometer, Karlsruhe, Germany) operating at 400 MHz (^{1}H-NMR) or 100 MHz (^{13}C-NMR) [30,31]. Fatty acid composition of the prepared methyl esters was studied through GCMS (QP2010SE SHIMADZU, Kyoto, Japan). GC-MS conditions are listed in Table 1. The complete procedure is presented in Supplementary Materials Table S4.

Table 1. Gas chromatograph conditions.

Parameter	Descriptions
Column	QP2010SE, Shimadzu, PEG-20M
	Length: 30 m, Internal diameter: 0.32 mm
	Film thickness: 1 um
Injector temperature	220 °C
Detector temperature (EI 250)	210 °C
Carrier gas	Helium, flow rate = 1.2 mL min^{-1}
Injection	V = 1 uL
Split ratio	Flow rate = 40:1
Temperature program	Initial temperature = 100 °C
	Rate of progression = 10 °C min^{-1}
	Final temperature = 210 °C, 20 min

2.6. Fuel Properties of KP Biodiesel

Once the pure form of biodiesel was obtained, its physico-chemical characterization was evaluated through American standard testing material (ASTM) and European norm (EN). Moreover, the oil characterization of KP biofuel was matched to those of petro-diesel. The physiochemical characterization contains kinematic viscosity at 40 °C (ASTM D445), density at 40 °C (ASTM D4052), flash point (ASTM D93), cloud point (ASTM D6751), pour point (ASTM D6751), acid value (ASTM D664), iodine value, oxidation stability at 110 °C (EN 14112), ash content (ASTM D874), saponification value, sulfur content (ASTM D5459), fire point and specific gravity were calculated and matched with standards. The higher heating value (HHV) of KP biodiesel was also resolute based on the eventual study [32].

3. Results and Discussion

To make biodiesel from inedible KP seed (Plant photographs are present in Figure 2a,b), we first extract the oil from the seed of KP, which has an oil content of 28–30% by Soxhlet extraction (calculated based on dry seeds), as presented in Supplementary Materials Table S1. Then transesterification occurs to covert the crude oil to biodiesel with methanol using KOH as a catalyst. When the reaction of transesterification was completed, the catalyst was removed by washing and centrifugation, and the final process mixture was put in a separatory glass funnel tube (Figure 2c). Two different layers appeared in a separatory glass funnel, the upper layers contain biodiesel, methanol, and glycerin, and the lower layer contains glycerides, methanol, glycerol, and catalyst. We carried out three to four times deionized hot water washing to remove the excess methanol and glycerol from biodiesel. Finally, dry cleaning was done with silica gel to remove the leftover water in the final biodiesel product (Figure 2d) [33].

Figure 2. (**a**) *Koelreuteria paniculata* plant (KP) (**b**) KP seed (**c**) Transesterification (the process of optimization) (**d**) Pure KP biodiesel after processing.

3.1. Koelreuteria Paniculata FAMEs Process of Optimizations

To obtain the maximum biodiesel yield, we studied different parameters of the transesterification process, i.e., methanol to oil molar ratio, KOH concentrations, temperature, time, and stirring rate. Detailed processes of the optimization were listed in Supplementary Materials Table S2. The initial yield of FAMEs was only 89%, which may have increased to 95.2% after the optimization. Five parameters were studied and analyzed under three or four different conditions to find out the optimal yield range of KP FAMES. The reason for using different parameters in different conditions is to check their best fit for the conditions to get the highest results. These five important variables and their analyses are explained respectively.

3.1.1. Effect of Methanol to Oil Molar Ratio

In both catalytic and non-catalytic reactions, the methanol/oil ratio is an important factor for the yield of methyl esters. Typically, transesterification (hydrolysis) needed 3 moles of alcohol for 1 mole of triglyceride to form 3 moles of fatty acid ester and 1 mole of glycerol. To obtained and evaluate the optimum molar ratio of methanol to oil, a range of ratios we have applied (4:1, 5:1, 6:1, and 7:1) in the experimentation (Figure 3a), while keeping other variables constant, such as temperature, stirring intensity and time. When the molar ratio was 6:1, the highest KPOB (KP oil biodiesel) yield was 94%. A stoichiometric ratio of 4:1 provides the lowest oil conversion to biodiesel (79%). It was further observed that if the methanol to oil ratio was lowered below the optimal range, no substantial conversion occurs in biodiesel production. The results showed that the overuse of methanol reduced FAMEs yield by 7:1 (86%); because of the additional use of methanol, the separation of esters and glycerol's becomes more complicated.

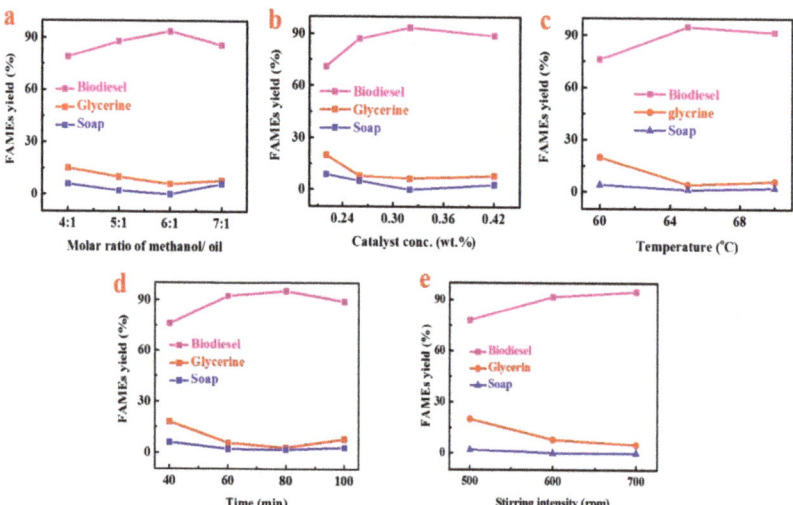

Figure 3. The effect of alcohol to oil molar ratio (**a**), KOH concentration (**b**), reaction temperature (**c**), time (**d**), and agitation rate (**e**) on KPOB production.

3.1.2. Impact of Catalyst Concentration on FAMEs Production

For the transesterification of triglycerides, various types of catalysts have been used, such as bases, acids, enzymes, or mixed catalysts. In the present study, KOH has been used as a base catalyst to convert the crude oil into fatty acid methyl ester. We achieved the highest biodiesel yield of 93.6%. The influence of catalyst concentration ratio on the FAMEs production was assessed (Figure 3b), whereas the additional operative settings of the transesterification process were kept constant. It was noted that the highest KP oil biodiesel (KPOB) product yield was achieved at a catalyst concentration of 0.32 wt.%. Additionally, it was witnessed that no important conversion occurred when the catalyst amount was higher than the optimal range.

3.1.3. Influence of Temperature on Yield

The reaction proportion was greatly affected by temperature. Usually, the reaction is carried out at a boiling point close to methanol (60–70 °C) at atmospheric pressure. The transesterification reaction was carried out at three different temperatures (60, 65, and 70 °C) depending on the oil type. The methanol boiling point is 64.7 °C. Temperatures higher than the boiling point will burn alcohol, resulting in lower yields. Correspondingly, to evaluate the optimum temperature range regarding the highest FAMEs production, the different reaction temperature was used to get maximum KPOB biodiesel yield during the present experiment. It is clear from the result (Figure 3c) that at 65 °C, maximum biodiesel production was achieved.

3.1.4. Influence of Reaction Time on FAMEs Yield

Transesterification reaction was carried out at an optimum reaction temperature (65 °C) for different periods from 40 min to 100 min. It was detected that the reaction time was an important parameter which had a direct influence on the product yield. To get a higher yield and fully convert crude oil to biodiesel, the reaction time played a very important role [34]. As shown in (Figure 3d), the biodiesel conversion and reaction time were directly proportional to each other. At 80 min, the highest biodiesel production (95.2%) was obtained, while at 40 min, the lowest biodiesel conversion was occurring (76%). It has been further observed from work in the experiment that after 80 min, the opposite reaction of hydrolysis started, and the FAMEs conversion ratio started decreasing, and longer reaction time resulted in a decrease in yield, which resulted in producing more soap and less FAMEs.

3.1.5. Influence of Agitation Rate on Biodiesel Yield

It was noticed through the transesterification process that the agitation rate could influence KPOB production, as shown in Figure 3e. The oil and fat were immiscible with the potassium hydroxide–methanol solution, so it was very important to mix by stirring during the transesterification reaction process. The reaction can become diffusion-controlled due to the slow rate of diffusion between the phases. Here we determined the direct relation between the KPOB production and the agitation strength, as with the increase in the stirring intensity, the biodiesel production increased. We evaluated that the 700 rpm was the optimum stirring rate for the higher production of biodiesel (95%) from KP seed oil to FAMEs. Though it was the same with the studies above [35] in which the authors concluded that a higher stirring intensity rate mixed the reaction well and gave the higher yield of methyl ester creations because elevated stirring strength promoted the homogenization of the reaction, which increased the yield of methyl ester.

3.2. KP FAMEs Characterizations

The physicochemical characterizations of the KP FAMEs were individually characterized (Table 2). The KP non-edible seed oil FAMEs had the following characteristics: viscosity 6.21 mm^2/s, density 0.879 g/cm^3, flashpoint 147 °C, ash content 0.002%, cloud point 2 °C and pour point −30 °C, flaming point 175 °C, saponification value 176.4 mg/g, specific gravity 0.88, and cold filter plugging point −18 °C.

Table 2. Physiochemical characterizations of KP seed oil biodiesel (KPSOB) in comparison with standards and plant fatty acid methyl esters (FAMEs).

Parameters	EN 14214	ASTM D-6751	Pedro-diesel	Present Results/KPOB	DPSOME [36]	SBOME [37]	DSOME [38]
Oil contents (wt.%)	–	–	–	28–30	10.50 ± 1.50	15–20	21.4
Density @ 15 °C (g/cm^3)	0.86–0.90	0.86–0.90	0.809	0.879	0.8781 ± 0.0012	–	0.882
Kinematic viscosity @ 40 °C (mm^2/s)	3.5–5.0	1.9–6.0	1.3–4.1	6.21	3.91 ± 0.25	4.50	4.33
Flashpoint (°C)	Min. 120	Min. 130	60–80	147	141	160	161.5
Ignition value	–	–	–	175	–	–	–
Acid value (mg KOH/g^{-1})	Max. 0.50	Max. 0.5	–	0.07	0.10 ± 0.02	0.15	0.10
Saponification value (mg KOH/g^{-1})	–	–	–	176.4	224	–	–
Iodine value (g I$_2$/100 mg)	Max. 120	Max. 120	–	80.7	48.51	–	–
Refractive index @ 20 °C	–	–	–	1.4901	1.4544	–	–
Cloud point (°C)	–	–	−15–5	2	3	4	–
Pour point (°C)	–	–	−2.0	−30	−1	−7	–
Cetane number	Min. 51	Min. 47	49.7	51	59.75	58.1	51.9
Free fatty acid (%)	–	–	–	0.91	–	–	–
HHV	–	–	–	23.39	39.52	–	–
Ash content	–	–	–	0.002	–	–	< 3
Specific gravity @15 (°C)	–	–	–	0.88	–	–	–
Cold filter plug point (°C)	Max. 19	Max. 19	−16	−18	–	–	−5
Sulphated ash content (wt.%)	Max. 0.02	–	–	0.003	–	0.008	0.004

3.3. Physical and Fuel Characterizations

The fuel and physicochemical properties of the produced KPOB were characterized by ASTM and EN standards. The results, as shown in Table 2, compared with the mentioned biofuel standards as ASTM D-6751 EN14214 and petrodiesel. The density of the KPBD at 40 °C was 0.879 g/cm^3, equivalent to the limits for petrodiesel (0.834 g/cm^3) and ASTM D-6751. The KPBD specific gravity value (0.88) determined in this analysis was comparable to the ASTM value of petro-diesel. The most imperative stuff of biodiesel was viscosity since it distressed the action of fuel inoculation apparatus, mostly at the temperatures less than 15 °C (viscosity affects fuel flow). The determined KPBD viscosity was 6.21 cps, somewhat higher than the ASTM D6751 rate of biodiesel and the ASTM rate of petroleum fuel, which needs further modification and improvement.

The pour point and cloud point dignified values for KPSO biodiesel KP seed oil (KPSO) biodiesel were −30 and 2 °C, respectively, which met the diesel fuel limits specified in ASTM (Table 2). Flashpoints are parameters deliberated in the management, packing, and protection of fuels and combustible ingredients. The detected KPSO biodiesel flash point (147 °C) was higher than petroleum diesel and was within the ASTM range. This shows that KPBD is more safe and secure than fossil fuels. Though, a higher flash point may result in a lower capability of the compression ignition engine to burn than petroleum diesel. The quantity of FFA in the biodiesel fuel indicates acid value. The quantity higher than 0.50 mg KOH g^{-1} may cause engine deterioration, 0.07 mg KOH g^{-1} is the identified KPBD acid value, which was expressively less than the ASTM standard diesel fuel (Table 2).

3.4. NMR Spectroscopy Analysis of Biodiesel

3.4.1. ^{1}H NMR Spectrum Analysis

The KP FAMEs were characterized by ^{1}H NMR; the KP ^{1}H NMR spectrum is presented in Figure 4a and Supplementary Materials Table S4. In the ^{1}H NMR spectrum, the terminal methyl proton (C-CH$_3$) appearance signals were detected between 0.88 and 0.90 ppm, and the aliphatic chain (-(CH$_2$) n-) associated signals were observed between 1.25 and 1.31 ppm. The β-methylene ester (CH$_2$-C-CO$_2$Me) bands appearance signals were detected around 1.58–1.65 ppm correspondingly, the methylene proton peaks near the base (-CH$_2$-C = C-) appear between 2.00 and 2.07 ppm attached to the allylic group. The methylene proton peaks signal were attributed between 2.27 and 2.31 ppm existing near the carbonyl group proton (-CH$_2$-COOMe), the existence of a methylene proton (-C = C-CH$_2$-C = C-) between the allylic groups associated between 2.75–2.80 ppm, and the single sharp peak at 3.59 ppm is expressed the ester bond (CH$_3$COO-CH) linked CH$_3$ group. The proton (-CH = CH-) from the glycerol moiety appears between 5.32–5.44 ppm.

3.4.2. ^{13}C NMR Spectrum Analysis

The KP biodiesel ^{13}C NMR spectrum is shown in Figure 4b and Supplementary Materials Table S5. A signal which shows the occurrence of an ester carbonyl carbon (-COO-) was observed at 174.20 ppm. The band's signals were observed in the spectrum between 127.84 and 130.19 ppm indicating the existence of unsaturation in the methyl ester. The occurrence of ester (C-O) methoxy carbon was observed at 51.38 ppm, due to the long carbon chain methylene carbon of the fatty acid methyl ester. The band's occurrence detected between 29.02 and 29.07 ppm, and the terminal carbon of the methyl group peaks was observed at 14.1 ppm, respectively.

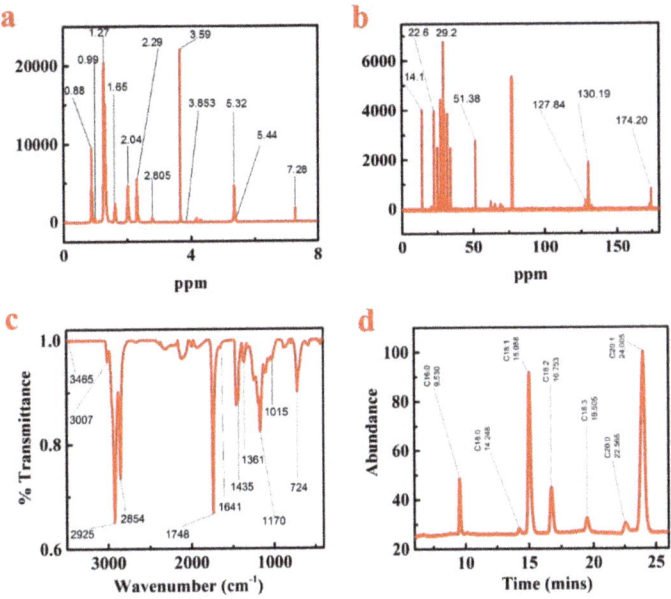

Figure 4. ^{1}H NMR (**a**), and ^{13}C NMR (**b**), KP (methyl ester) FTIR study (**c**), KP seed oil biodiesel GC–MS fatty acid compositions analysis (**d**).

3.5. KP FAMEs FT-IR Spectroscopy Analysis

To identify the functional groups and the bands corresponding to various stretching and bending vibrations in KP biodiesel samples, the FT-IR spectroscopy of the mid-infrared region has been used, as presented in Figure 4c and the Supplementary Materials Table S3. The two resilient ester representative absorption bands were detected from carbonyl (νC = O) around 1750–1730 cm^{-1} and C-O at 1300–1000 cm^{-1} [39]. The stretching vibrations and bending vibrations (ρCH$_2$) of CH$_3$, CH$_2$, and CH appeared at 2980–2950, 2950–2850, 3050–3000 cm^{-1}, and at 1475–1350, 1350–1150, 722 cm^{-1}, correspondingly [40]. The absorption peaks of the sample were detected in KP biodiesel to be 3465, 3007, 2925, 2854, 1748, 1641, 1435, 1361, 1175, 1015, and 724 cm^{-1}, respectively. The peaks presence in KP FAMEs at 1435 and 1170 cm^{-1} specifies the conversion of crude oil to biodiesel. Figure 4c shows the KPME FTIR spectrum, and the strong absorption peak at 2854 cm^{-1} and 2925 cm^{-1} is just because of the alkane group of C-H stretching vibration. The C-H bending vibration appears at 1435 cm^{-1} due to strong absorption. All of the single bands symbolize saturated functional groups. The C=O stretching frequency peak at 1748 cm^{-1} is due to strong absorption, which is composed of an unsaturated functional group and is called an ester. Further, due to the C-O stretching vibration of the ester, the strong band appeared at 1015 cm^{-1} and 1170 cm^{-1}.

3.6. GC/MS Analysis

The biodiesel of KP crude oil modified by methyl ester was evaluated by gas chromatography and mass spectrometry. The GCMS spectral data are shown in Figure 4d. The NIST-14 library matching software was used to identify the peaks. After evaluation, each single peak was matched with fatty acid methyl ester [18]. The retention time and position of the determined peaks are presented in Table 3.

Table 3. *Koelreuteria paniculata* seed oil biodiesel fatty acid composition.

S/No	Fatty Acids	Retention Time	Number of Carbons	Fatty Acids (%)	Chemical Name	Chemical Structure	Molecular Weight
1	Palmitic acid	9.530	C16:0	9.7	Hexadecanoic acid, methyl ester		270
2	Stearic acid	14.248	C18:0	1.8	Methyl stearate		298
3	Oleic acid	15.058	C18:1	25.5	9-Octadecenoic acid (Z)-, methyl Ester		296
4	Linoleic acid	16.753	C18:2	8.5	9,12-Octadecadienoic acid (Z, Z)-, methyl ester		294
5	α-Linolenic acid	19.505	C18:3	3.6	α-Linolenic acid		292
6	Arachidic acid	22.565	C20:0	2.4	Eicosanoic acid, methyl ester		326
7	Gondoic acid	23.005	C20:1	48.5	Cis-11-Eicosenoic acid, methyl ester		324

3.7. ICP-OES and Elemental Analyzer (EA) Study

The elements present in fuels are harmful because they cause many difficulties, such as promoting fuel degradation, corrosion of machinery, operational problems, environmental pollution and other social problems, as well as harmful effects on health [41]. The fundamentals whose quantity in biodiesel requires to be controlled are sodium (Na) and potassium (K) are important limiting quantities in the production mechanism of biodiesel, while phosphorus (P) comes from raw materials and its controlling quantities are very important. The element concentration of *Koelreuteria pniculata* oil biodiesel (KPOBD) and HSD (high speed diesel) was compared. KP results confirm that the concentration of many base metals in KPOBD is lower than that in HSD. Elements such as Ca, Mg, Na and K in biodiesel directly enter the injector, and locomotive sediments stimulate the drainage and piston and pass through the filter plug [42]. K, Na, Mg, and Ca are present in Supplementary Materials Table S6. The KPOBD (6.14, 544.9, 32.10, 14.90 µg/g) was lower as compared to HSD (213.3, 868.3, 35.6 and 21.4 µg/g). The highest acceptable concentration of Na and K in biodiesel was 5 mg kg^{-1}, while P was 10 mg kg^{-1} (Figure 5).

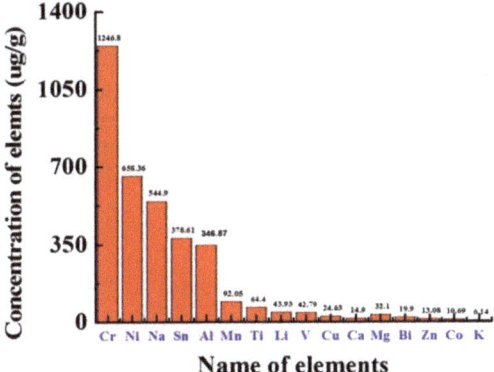

Figure 5. KPBD ICP-OES elemental analysis.

A detailed discussion of N, H, C, and O is shown in Table 4 and Supplementary Materials Figure S1. The petrodiesel has no oxygen, while biodiesel has oxygen, which is the main difference between biodiesel and petrodiesel. The main advantage of the presence of oxygen is that it can reduce ignition delay, increase combustion, and come entirely from fuel combustion, a purposeful cause that will reduce exhaust emissions, CO, and particulate matter. Therefore, several scientists [43,44] have proposed that biodiesel with higher oxygen content can significantly reduce particulate emissions from fuel engines. Table 4 specifies KP B100 as 12%, whereas KP B100 for most biodiesel is approximately 10% [45,46]. The higher hydrogen content (12.73%) of biodiesel utilization makes KP biodiesel more attractive [47]. In gaseous and liquid fuels, hydrogen burns faster and hydrogen carbon is higher, which means the fuel has a higher molecular content of hydrogen, better combustion capacity and cleaner. Since hydrogen has a higher calorific value than carbon, the calorific value must increase as the ratio of hydrogen to carbon atoms increases.

Table 4. Elemental analyzer study of KPBD and their comparison with other plant biodiesel.

Ultimate Analysis	KP-BD	Pistachio Shell [48]	Peach Stones [49]	Apricot Kernel Shells [50]	Cherry Stones [51]	Mahua Seed [52]
C%	72.54	42.41	45.92	47.33	52.48	61.24
H%	12.73	5.64	6.09	6.37	7.58	8.40
N%	2.73	0.070	0.580	0.370	4.54	4.12
O%	12	51.87	47.38	45.93	35.30	25.50
HHV (MJ/kg)	23.39	22.21	24.07	24.29	24.11	25.30

4. Conclusions

A novel plant seed of *Koelreuteria paniculata* was explored to produce biodiesel effectively with a simple, cheap, and energy-saving base-catalyzed methanol transesterification process. Through the optimization transesterification conditions: oil to methanol ratio (6:1), catalyst concentration (0.32 g wt.%), temperature (65 °C), stirring rate (700 rpm) and time (80 min), the highest biodiesel yield of 95.2% was obtained, which is favorable for commercial application. The physico-chemical characterizations of the biodiesel, i.e., flash point, pour point, cloud point, and density, are within the ASTM (D6751) and (EN14214) standards, the values of fuel properties were found to be comparable with mineral diesel, which is good for engine application. FTIR, NMR, and GC-MS analysis confirmed the complete conversion of crude oil to FAMEs. Finally, we used elements analysis ensured the environmentally friendly of biodiesel. Based on these analyses, it is stated that the biodiesel derived from KP seeds is considered to be an alternative and renewable source of fuel for energy production, which could be a promising substitute for petroleum diesel.

Supplementary Materials: The following are available online at http://www.mdpi.com/1996-1073/13/4/791/s1, Figure S1: KP plant Description, Figure S2: Mechanical extraction of seed, Figure S3: ICP-OES and EA study of KP biodiesel for elemental analysis. Figure S4: GC-MS Procedure, Table S1: Source collection, Oil extraction and transesterification of non-edible KP seed oil as biofuel, Table S2: KP FAMEs process of optimization, Table S3: FTIR data presenting various functional groups in KP FAMEs, Table S4: ^{1}H NMR spectroscopic data showing the chemical composition of various methyl esters (Methoxy proton) in KP biodiesel (FAMEs), Table S5: ^{13}C NMR spectroscopic data showing the chemical shift values corresponding to various structural features in KP (Methoxy carbon) FAMEs, Table S6: Shows KPBD ICP-OES elements concentration (ug/g) in comparison with petro-diesel.

Author Contributions: J.C. and Z.Y. supervised the research and revised the manuscript, Chen Hang help in seed collection and revised the manuscript, I.U.K., done the experiment and write the manuscript. All authors have read and agreed to the published version of the manuscript.

Funding: This work was supported by the National Natural Science Foundation of China (NSFC, 21801048), the National Programs for Nano-Key Project (2017YFA0206700), the National Natural Science Foundation of China (21835004), 111 Project from the Ministry of Education of China (B12015) and the Fundamental Research Funds for the Central Universities, Nankai University (63191711 and 63191416).

Conflicts of Interest: The authors declare no conflict of interest.

References

1. Röllin, H. Evidence for health effects of early life exposure to indoor air pollutants: What we know and what can be done. *J. Clean Air* **2017**, *27*, 2–3. [CrossRef]
2. Sharma, Y.C.; Singh, B.; Upadhyay, S.N. Advancements in development and characterization of biodiesel. A review. *Fuel* **2008**, *87*, 2355–2377. [CrossRef]
3. Xue, B.J.; Luo, J.; Zhang, F.; Fang, Z. Biodiesel production from soybean andJatropha oils by magnetic CaFe$_2$O$_4$-Ca$_2$Fe$_2$O$_5$-based catalyst. *Energy* **2014**, *68*, 584–591. [CrossRef]
4. Samadi, S.; Karimi, K.; Behnam, S. Simultaneous biosorption and bioethanol production from lead-contaminated media by mucor indicus. *J. Biofuel Res.* **2017**, *4*, 545–550. [CrossRef]
5. McKendry, P. Energy production from biomass conversion technologies. *Bioresour. Technol. Part A* **2002**, *83*, 47–54. [CrossRef]

6. Warren Spring Laboratory. *Fundamental Research on the Thermal Treatment of Wastes and Biomass: Literature Review of Part Research on Thermal Treatment of Biomass and Waste*; B/T1/00208/Rep/1; ETSU: Johnson City, TN, USA, 1993.
7. Khan, I.U.; Yan, Z.H.; Chen, J. Optimization, transesterification and analytical study of Rhus typhina non-edible seed oil as biodiesel production. *Energies* **2019**, *12*, 4290.
8. Guedes, R.E.; Luna, A.S.; Torres, A.R. Operating parameters for bio-oil production in biomass pyrolysis: A review. *J. Anal. Appl. Pyrol.* **2018**, *129*, 134–149. [CrossRef]
9. Mucak, A.; Karabektas, M.; Hasimoglu, C.; Ergen, G. Performance and emission characteristics of a diesel engine fuelled with emulsified biodiesel-diesel fuel blends. *Int. J. Autom. Eng. Technol.* **2016**, *5*, 176–185. [CrossRef]
10. Chew, K.W.; Yap, J.Y.; Show, P.L.; Suan, N.H.; Juan, J.C.; Ling, T.C.; Lee, D.J.; Chang, J.S. Microalgae biorefinery: High value products perspectives. *Bioresour. Technol.* **2017**, *229*, 53–62. [CrossRef]
11. Martín-Lara, M.A.; Ortuño, N.; Conesa, J.A. Volatile and semivolatile emissions from the pyrolysis of almond shell loaded with heavy metals. *Sci. Total Environ.* **2018**, *613–614*, 418–427.
12. Tremblay, A.Y.; Montpetit, A. The in-process removal of sterol glycosides by ultrafiltration in biodiesel production. *J. Biofuel Res.* **2017**, *4*, 559–564. [CrossRef]
13. Rashed, M.M.; Kalam, M.A.; Masjuki, h.H.; Mofijur, M.; Rasul, M.G. Performance and emission characteristics of a diesel engine fueled with palmjatropha, and moringa oil methyl ester. *Ind. Crops Prod.* **2016**, *79*, 70–76. [CrossRef]
14. Gui, M.M.; Lee, K.T.; Bhatia, S. Feasibility of edible oil vs. non-edible oil vs. waste edible oil as biodiesel feedstock. *Energy* **2008**, *33*, 1646–1653. [CrossRef]
15. Atadashi, I.M.; Aroua, M.K.; Abdul Aziz, A.R.; Sulaiman, N.M.N. Production of biodiesel using high free fatty acid feedstocks. *Renew. Sustain. Energy Rev.* **2012**, *16*, 3275–3285. [CrossRef]
16. Shafiei, A.; Rastegari, H.; Ghaziaskar, H.S.; Yalpani, M. Glycerol transesterification with ethyl acetate to synthesize acetins using ethyl acetate as reactant and entrainer. *J. Biofuel Res.* **2017**, *4*, 565–570. [CrossRef]
17. Gandhi, M.; Ramu, N.; Bakkiya Raj, S. Methyl ester production from Schlichera oleosa. *Int. J. Pharm. Sci. Res.* **2011**, *2*, 1244–1250.
18. Wang, L.B.; Yu, H.Y. Biodiesel from Siberian apricot (Prunus sibirica L.) seed kernel oil. *Bioresour. Technol.* **2012**, *112*, 335–338. [CrossRef] [PubMed]
19. Silitonga, A.S.; Masjuki, H.H.; Mahlia, T.M.I.; Ong, H.C.; Atabani, A.E.; Chong, W.T. A global comparative review of biodiesel production from jatropha curcas using different homogeneous acid and alkaline catalysts: Study of physical and chemical properties. *Renew. Sustain. Energy Rev.* **2013**, *24*, 514–533. [CrossRef]
20. Ashraful, A.M.; Masjuki, H.H.; Kalam, M.A.; Rizwanul Fattah, I.M.; Imtenan, S.; Shahir, S.A.; Mobarak, H.M. Production and comparison of fuel properties, engine performance, and emission characteristics of biodiesel from various non-edible vegetable oils: A review. *Energy Convers. Manag.* **2014**, *80*, 202–228. [CrossRef]
21. Aliyu, B.; Agnew, B.; Douglas, S. Croton megalocarpus (Musine) seeds as potential source of bio-diesel. *Biomass Bioenergy* **2010**, *34*, 1495–1499. [CrossRef]
22. Atapour, M.; Kariminia, H.R. Characterization and transesterification of Iranian bitter almond oil for biodiesel production. *Appl. Energy* **2011**, *88*, 2377–2381. [CrossRef]
23. Nehdi, I.A.; Sbihi, H.M.; Al-Resayes, S.I. Rhazya stricta Decne seed oil as an alternative, non-conventional feedstock for biodiesel production. *Energy Convers. Manag.* **2014**, *81*, 400–406. [CrossRef]
24. Reshad, A.S.; Tiwari, P.; Goud, V.V. Extraction of oil from rubber seeds for biodiesel application: Optimization of parameters. *Fuel* **2015**, *150*, 636–644. [CrossRef]
25. Fadhil, A.B.; Aziz, A.M.; Altamer, M.H. Biodiesel production from Silybum marianum L. seed oil with high FFA content using sulfonated carbon catalyst for esterification and base catalyst for transesterification. *Energy Convers. Manag.* **2016**, *108*, 255–265. [CrossRef]
26. Aldobouni, I.A.; Fadhil, A.B.; Saied, I.K. Optimized alkali catalyzed transesterification of wild mustard (Brassica juncea L.) seed oil. *Energy Sources Part A* **2016**, *38*, 2319–2325. [CrossRef]
27. Chen, Y.H.; Chen, J.H.; Luo, Y.M. Complementary biodiesel combination from tung and medium-chain fatty acid oils. *Renew. Energy* **2012**, *44*, 305–310. [CrossRef]

28. Kpikpi, W.M. Jatropha curcus as a vegetable source of renewable energy. ANSTI Sub. Network Meeting on Renew. *Energy* **2002**, *3*, 18–22.
29. Ahmad, M.; Khan, M.A.; Zafar, M.; Sultana, S. *Practical Handbook on Biodiesel Production and Properties*; Taylor and Francis: Boca Raton, FL, USA, 2012; pp. 1–157.
30. Bhandari, D.C.; Chandel, K.P.S. Status of rocket germplasm in India: Research accomplishments and priorities. In *Rocket: A Mediterranean Crop for the World*; International Plant Genetics Research Institute: Legnaro, Italy, 1996; Volume 67, pp. 13–14.
31. Christie, W.W. *Lipid Analysis*, 3rd ed.; Oily Press: Bridgwater, UK, 2003.
32. Sen, N.; Kar, Y. Pyrolysis of black cumin seed cake in a fixed-bed reactor. *Biomass Bioenerg.* **2011**, *3*, 4297–4304. [CrossRef]
33. Usta, N.; Öztürk, E.; Can, O.; Conkur, E.S.; Nas, S.; Çon, A.H. Combustion of biodiesel fuel produced from hazelnut soapstock/waste sunflower oil mixture in a diesel engine. *Energy Convers. Manag.* **2005**, *46*, 741–755. [CrossRef]
34. Kafuku, G.; Mbarawa, M. Effects of biodiesel blending with fossil fuel on flow properties of biodiesel produced from non-edible oils. *Int. J. Green Energy* **2010**, *7*, 434–444. [CrossRef]
35. Vedaraman, N.; Puhan, S.; Nagarajan, G.; Velappan, K.C. Preparation of palm oil biodiesel 305 and effect of various additives on NOx emission reduction in B20: An experimental study. *Int. J. Green Energy* **2011**, *8*, 383–397. [CrossRef]
36. Abdelrahman, B.F.; Mohammed, A.A.; Liqaa, I.S. Date (Phoenix dactylifera L.) palm stones as a potential new feedstock for liquid bio-fuels production. *Fuel* **2017**, *210*, 165–176.
37. Karmakar, A.; Karmakar, S.; Mukherjee, S. Properties of various plants and animal feedstock's for biodiesel production. *Bioresour. Technol.* **2010**, *101*, 7201–7210. [CrossRef] [PubMed]
38. Wang, R.; Zhou, W.W.; Hanna, M.A.; Zhang, Y.P.; Bhadury, P.S.; Wang, Y.; Song, B.A.; Yang, S. Biodiesel preparation, optimization, and fuel properties from non-edible feedstock, Datura stramonium L. *Fuel* **2012**, *91*, 182–186. [CrossRef]
39. Soares, I.P.; Rezende, T.F.; Silva, R.C.; Castro, E.V.R.; Fortes, I.C.P. Multivariate calibration by variable selection for blends of raw soybean oil/biodiesel from different sources using Fourier Transform Infrared Spectroscopy (FT-IR) spectra data. *Energy Fuels* **2008**, *22*, 2079–2083. [CrossRef]
40. Safar, M.; Bertrand, D.; Robert, P.; Devaux, M.F.; Genut, C. Characterization of edible oils, butters and margarines by Fouier Transform Infrared Spectroscopy with attenuated total reflectance. *J. Am. Oil Chem. Soc.* **1994**, *71*, 371–377. [CrossRef]
41. Schober, S.; Mittelbach, M. Influence of Diesel Particulate Filter Additives on Biodiesel Quality. *Eur. J. Lipid Sci. Technol.* **2005**, *107*, 268–271. [CrossRef]
42. McCormick, R.L.; Alleman, T.L.; Ratcliff, M.; Moens, L.; Lawrence, R. Survey of the Quality and stability of Biodiesel and Biodiesel Blends in the United States in 2004. In *Technical Report of National Renewable Energy Laboratory*; National Renewable Energy Laboratory (NREL): Golden, CO, USA, 2005.
43. Song, H.; Quinton, K.S.; Peng, Z.; Zhao, H.; Ladommatos, N. Effects of Oxygen Content of Fuels on Combustion and Emissions of Diesel Engines. *Energies* **2016**, *9*, 28. [CrossRef]
44. Mwang, J.K.; Lee, W.J.; Chang, Y.C.; Chen, C.Y.; Wang, L.C. An overview: Energy saving and pollution reduction by using green fuel blends in diesel engines. *Appl. Energy* **2015**, *159*, 214–236. [CrossRef]
45. Singh, D.; Subramanian, K.A.; Juneja, M.; Singh, K.; Singh, S. Investigating the effect of fuel cetane number, oxygen content, fuel density, and engine operating variables on NOx emissions of a heavy duty diesel engine. *Environ. Prog. Sustain. Energy* **2017**, *36*, 214–221. [CrossRef]
46. Demirbas, A. Combustion Efficiency Impacts of Biofuels. *Energy Sources Part A* **2009**, *31*, 602–609. [CrossRef]
47. Huber, G.W.; Iborra, S.; Corma, A. Synthesis of transportation fuels from biomass: Chemistry, catalysts, and engineering. *Chem. Rev.* **2006**, *106*, 4044–4098. [CrossRef] [PubMed]
48. Acıkalın, K.; Karaca, F.; Bolat, E. Pyrolysis of pistachio shell: Effects of pyrolysis conditions and analysis of products. *Fuel* **2012**, *95*, 169–717. [CrossRef]
49. Uysal, T.; Duman, G.; Onal, Y.; Yasa, I.; Yanik, J. Production of activated carbon and fungicidal oil from peach stone by two-stage process. *J. Anal. Appl. Pyrol.* **2014**, *108*, 47–55. [CrossRef]
50. Demiral, I.; Kul, S.C. Pyrolysis of apricot kernel shell in a fixed-bed reactor: Characterization of bio-oil and char. *J. Anal. Appl. Pyrol.* **2014**, *107*, 17–24. [CrossRef]

51. Duman, G.; Okutucu, C.; Ucar, S.; Stah, R.; Yanik, J. The slow and fast pyrolysis of cherry seed. *Bioresour. Technol.* **2011**, *102*, 1869–1878. [CrossRef]
52. Pradhan, D.; Singh, R.K.; Bendu, H.; Mund, R. Pyrolysis of Mahua seed (Madhuca indica)—Production of biofuel and its characterization. *Energy Convers. Manag.* **2016**, *108*, 529–538. [CrossRef]

© 2020 by the authors. Licensee MDPI, Basel, Switzerland. This article is an open access article distributed under the terms and conditions of the Creative Commons Attribution (CC BY) license (http://creativecommons.org/licenses/by/4.0/).

Article

Isolation and Characterization of Two Microalgal Isolates from Vietnam with Potential for Food, Feed, and Biodiesel Production

Thao Nguyen Luu [1,2], Zouheir Alsafra [3], Amélie Corato [4], Daniele Corsaro [5], Hung Anh Le [6], Gauthier Eppe [3,*] and Claire Remacle [1,*]

1. Genetics and Physiology of Microalgae, InBios/Phytosystems Research Unit, University of Liege, 4000 Liege, Belgium; luuthaonguyen@iuh.edu.vn
2. Institute of Biotechnology and Food Technology, Industrial University of Ho Chi minh City, 71406 Ho Chi minh, Vietnam
3. Laboratory of Mass Spectrometry, MolSys Research Unit, University of Liege, 4000 Liege, Belgium; zouheir.alsafra@doct.ulg.ac.be
4. Bioenergetics, InBios/Phytosystems Research Unit, University of Liege, 4000 Liege, Belgium; Amelie.Corato@uliege.be
5. CHLAREAS, 12, rue du Maconnais, F-54500 Vandoeuvre-lès-Nancy, France; corsaro@gmx.fr
6. Institute of Environmental Science, Engineering and Management, Industrial University of Ho Chi minh City, 71406 Ho Chi minh, Vietnam; lehunganh@iuh.edu.vn
* Correspondence: g.eppe@uliege.be (G.E.); c.remacle@uliege.be (C.R.); Tel.: +32-366-34-22 (G.E.); +32-366-38-12 (C.R.)

Received: 29 December 2019; Accepted: 11 February 2020; Published: 18 February 2020

Abstract: Microalgae are promising feedstock for the production of biodiesel and diverse medium- and high-value products such as pigments and polyunsaturated fatty acids. The importance of strain selection adapted to specific environments is important for economical purposes. We characterize here two microalgal strains, isolated from wastewater of shrimp cultivation ponds in Vietnam. Based on the 18S rDNA-ITS region, one strain belongs to the Eustigmatophyceae class and is identical to the *Nannochloropsis salina* isolate D12 (JX185299.1), while the other is a Chlorophyceae belonging to the *Desmodesmus* genus, which possesses a S516 group I intron in its 18S rDNA gene. The *N. salina* strain is a marine and oleaginous microalga (40% of dry weight (DW) at stationary phase) whole oil is rich in saturated fatty acids (around 45% of C16:0) suitable for biodiesel and contains a few percent of eicosapentaenoic acid (C20:5). The *Desmodesmus* isolate can assimilate acetate and ammonium and is rich in lutein. Its oil contains around 40%–50% α-linolenic acid (C18:3), an essential fatty acid. Since they tolerate various salinities (10% to 35‰), both strains are thus interesting for biodiesel or aquaculture valorization in coastal and tropical climate where water, nutrient, and salinity availability vary greatly depending on the season.

Keywords: microalga; fatty acid; Vietnam; *Nannochloropsis*; *Desmodesmus*

1. Introduction

Microalgae are phototrophic organisms representing a promising feedstock for the production of biodiesel and diverse medium- and high-value products [1]. They can live in various habitats, including freshwater, brackish water, and marine environments such as oceans, lagoons, and ponds where the salinity can be subject to fluctuations due to drought or heavy rains in tropical climates. The importance of strain selection for economically viable algal-based bioproducts is recognized [2] and the selection should focus on different criteria such as the capacity to accumulate large amounts of the desired compound or to grow in specific conditions (salt tolerance, temperature, etc.) [2]. In addition, utilization of the whole microalgal biomass is desirable to increase the chances of commercial success. In this context, coupling production of value-added bioproducts with microalgal biofuel has been suggested to be a promising technology to reduce costs [3].

Medium- and high-value products include long chain polyunsaturated fatty acids (PUFAs) and carotenoids. ω-3 PUFAs [eicosapentaenoic acid (EPA C20:5n−3) and docosapentaenoic acid (DHA C22:6n−3)] have nutritive values important for aquaculture and for human health [4,5]. Nowadays, EPA and DHA are mainly found in fish oil, but overfishing has reduced wild fish stocks [6]. Thus, industries look for additional sources of natural EPA and DHA and microalgae that contain substantial amount of ω-3 PUFAs seem to be interesting. Carotenoids are pigments that are found important as they protect cells from oxidative damage. They represent food and feed additives and health-promoting supplements [7]. Microalgae containing substantial amounts of pigments such as β-carotene, lutein, canthaxanthin, astaxanthin, phycocyanin, and fucoxanthin are thus commercially interesting [4,8].

With the aim to find strains suitable for growth in tropical climate and adapted to changing salinity environment, we have isolated two microalgae from wastewater of shrimp cultivation ponds located in the coastal region of the Ninh Thuan province, Vietnam. The two strains were identified based on the 18S rDNA-ITS region. One strain belongs to the Eustigmatophyceae class and is identical to the *Nannochloropsis salina* isolate D12 (JX185299.1) recorded in GenBank while the other is a Chlorophyceae belonging to the *Desmodesmus* genus. They are part of distant phyla. *Desmodesmus* belongs to Chlorophyta characterized by chloroplasts derived from primary endosymbiosis with a cyanobacterium whose most probable living counterpart is the fresh water-dwelling *Gloeomargarita lithophora*. *Nannochloropis* belongs to Stramenopiles characterized by chloroplasts derived from a secondary symbiosis event with a red alga [9–12]. Characterization of both strains is performed in terms of growth, protein content, fatty acid, pigment profile, and resistance to salinity stress in order to assess their potential for biotechnological applications in this region.

2. Materials and Methods

2.1. Isolation and Purification of Microalgae

Marine water samples were collected from shrimp cultivation ponds at Marine seed center at level I, Ninh Thuan province, Vietnam (Latitude 11°30'33.4" N—Longitude 109°00'35.6" E) and first enriched in the laboratory using liquid F/2 medium with salinity of 24‰. The enrichment culture was aerated using an orbital shaker. The growth was maintained at 35 ± 1 °C under constant illumination. After 2 weeks, 100 µL of greenish enrichment cultures were spread on plate of solid F/2 medium (salinity of 24‰) containing 1.5% agar and ampicillin (0.1 µg mL^{-1}). Ampicillin was used to eliminate the bacterial contaminants. The solid plates were incubated under constant illumination at 22 ± 1 °C until microalgal colonies along with bacterial and fungal colonies appeared on plates. Streak plate technique was repeated until the axenic cultures were obtained. To ensure the axenicity of isolates, light microscope and lab binocular were used to observe the algal cells and colonies, respectively.

2.2. DNA Isolation, PCR and DNA Sequencing

DNA isolation was performed according to [13]. The 18S rDNA–ITS sequence of isolates *nl3* and *nl6* was amplified using primers NS1 and ITS4 with standard PCR protocol and the PCR product was sequenced using these two primers, as well as the internal primers NS2, NS3, NL4, NS5, NS8, and ITS1 (Beckman Coulter Company, Takeley, UK).

The overlapping partial sequences between two consecutive sequences were assembled using NCBI Blast Tool (Standard nucleotide blast) to obtain a complete 18S rDNA–ITS1-5.8S-ITS2 sequence which was compared with sequences on NCBI database for hits. The sequence of *nl3* is deposited in GenBank (MN746324).

2.3. Microalgal Growth

Isolate *nl3* and *nl6* were cultivated in 250 mL Erlenmeyer flasks containing sterile TAP (Tris-Acetate-Phosphate) medium [14] or Guillard's F/2 medium (Sigma-Aldrich, St Louis, MO, USA) [15] with salinity ranging between 0–35‰ when specified. Cultures were aerated on a shaker and continuously illuminated with constant illumination of 200 µmol m^{-2}s^{-1} at 25 ± 1 °C. The optical density at wavelength of 750 nm was measured every day to evaluate the cell density [16] using three biological replicates. The data were expressed as mean ± standard deviation (±SD).

$$\mu = (\ln N - \ln N_0)/(t - t_0) \tag{1}$$

The specific growth rate (μ) was calculated using Formula (1) [17] in which N_0 and N stand for the OD$_{750}$ value at the beginning (t_0) and the end of exponential phase (t), respectively.

2.4. Biomass Analysis

2.4.1. Dry Weight

40 mL of culture was harvested in a falcon tube. The tube was centrifuged at 1500 g for 10 min. The supernatant was discarded. Three mL of distilled water were added to the pellet and transferred to a pre-weighed glass tube. The glass tube was centrifuged at 800 g for 10 min to remove the supernatant. The glass tube containing wet biomass was allowed to dry in oven at 80 °C until constant weight was achieved. Three biological triplicates were used. The data were expressed as mean ± standard deviation (±SD).

2.4.2. Fatty Acid Methyl Ester Analysis (FAME) Analysis

FAME preparation was carried out using a process involving two main steps: lipid extraction and transesterification. Two to ten mL aliquot of each algal culture was harvested in a falcon tube and centrifuged for 10 min at 4000× g, 4 °C. Two mL of Folch reagent (Chloroform–Methanol) were added to the pellet. The suspension was sonicated in an ice batch for 10 min to break the cell wall. One mL of the cell/Folch mixture was transferred to a sterile glass tube using Pasteur pipette. Thirty µL of internal standard C15:0 were added to the tube using a Hamilton syringe prior to transesterification reaction. The mixture was evaporated to dryness at 80 °C. One mL of BF$_3$/Methanol (Sigma-Aldrich, St Louis, MO, USA), was added in the tube, sealed and homogenized. The mixture was heated at 95 °C for 10 min in a water bath to conduct transesterification reaction. The reaction tube was slowly cooled down to room temperature within 15–20 min. One mL of water and 1 mL of hexane were added and the reaction tube was stirred for 30 min on a carousel. A centrifugation at 200× g for 10 min, 4 °C was then conducted. The lower aqueous phase was eliminated from the tube using a Pasteur pipette. Again, 1 mL of ultra-pure water was poured in the tube, mixed, and centrifuged for 10 min at 200× g, 4 °C. The tube was stored in the freezer overnight. The organic phase (top phase) containing FAMEs was transferred to a two-milliliter chromacol vial, hermetically sealed with a cap and stored at -20 °C until analysis.

FAME analyses were conducted on a Trace GC2000-PolarisQ ion trap mass spectrometer (ThermoScientific, Waltham, MA, USA) coupled with CTC Combi-Pal autosampler (CTC Analytics, Zwingen, Switzerland). FAMES were separated in GC column SP2331 (30 m × 0.25 mm ID × 0.20 µm film thickness, Supelco, Bellefonte, PA, USA). with helium as carrier gas. The GC temperature program consisted of 4 thermal steps; the column was initially held at 60 °C for 2 min; then ramped to 180 °C at a rate of 5 °C min^{-1}; followed by a rate of 10 °C min^{-1} to 250 °C maintained during 1 min. Mass spectral profile, comparison to external standards and retention time were criteria used to identify FAMES. Quantification of FAMEs was done using calibration curves of FAME external standards and C15:0 internal standard. Three biological triplicates were used unless otherwise stated, and the data were expressed as mean ± standard deviation (±SD).

2.4.3. Biodiesel Fuel Properties Based on FAME Profiles

Some properties of biodiesel were calculated using the following equations [18]

$$\text{Average degree of unsaturation (ADU)} = \sum A \times k$$

A: % of each fatty acid on total fatty acids; k: number of double bonds in each fatty acid Kinematic viscosity = $-0.6316 \times \text{ADU} + 5.2065$ (kV) (mm^2.s^{-1}).

Kinematic viscosity (kV) = $-0.6316 \times \text{ADU} + 5.2065$ (mm^2 s^{-1})
Specific gravity (SG) = $0.0055 \times \text{ADU} + 0.8726$ (kg L^{-1})
Cloud point (CP) = $-13.356 \times \text{ADU} + 19.994$ (°C)
Cetane number (CN) = $-6.6684 \times \text{ADU} + 62.876$
Iodine value (IV) = $74.373 \times \text{ADU} + 12.71$ (gI$_2$ 100 g^{-1})
Higher heating value (HHV) = $1.7601 \times \text{ADU} + 38.534$ (Mj kg^{-1})

2.4.4. Protein Content

1 mL of culture was harvested for protein assay. The pellet containing wet biomass was suspended in 1 mL 50 mm HEPES-KOH pH 7.2 and 5 µL phenylmethanesulfonyl fluoride (PMSF) 100 µM. About 600 mg of glass beads (0.5 mm) were added to the tube and vortexed for 40 min at maximum level, 4 °C. A 20 µL aliquot of the extract was used for Bradford assay [19]. The sample volume was brought to 100 µL with buffer solution (0.1 M NaOH, 0.1% Triton). The sample was mixed with 1 mL of Bradford reagent (1X). Dye will bind to protein to have a blue protein–dye form. Optical absorbance of mixture was read by PDA UV/VIS Lambda 265 Spectrophotometer (PerkinElmer, Waltham, MA, USA) at 595 nm. The concentration of protein in sample was presented as µg/mL and calculated based on the standard curve using optical absorbance against bovine serum albumin (BSA) concentration. Three biological triplicates were used unless otherwise stated. The data were expressed as mean ± standard deviation (±SD).

2.4.5. Pigment Analysis

1 mL of culture was harvested to collect the biomass at 4 °C. Pigments were extracted in 1 mL CH$_2$Cl$_2$:CH$_3$OH (1:3) using glass beads of 0.5 mm at 4 °C for 40 min. The supernatant was collected by centrifugation at 20,000× g for 15 min and filtered through 0.2 um pores before loading into HPLC system. The HPLC device was a Shimadzu equipped with a photodiode array detector DAD (SPD-M20A). The pigment separation was conducted in reverse phase gradient mode with a CORTECS C18 column (90 Å pore size, 2.7 µm solid-core particles, 4.6 mm × 150 mm, 3/pk, Waters), under a flow of 1 mL min^{-1}, at 25 °C. The solvents used were NH$_4$Ac 50 mm + 75% Acetonitrile (A), 90% Acetonitrile (B) and Ethyl Acetate (100%) (C), and the gradient was 0 min—100% A; 2.5 min—100% B; 3.1 min—90% B + 10% C; 8.1 min—65% B + 35% C; 13.5 min—40% B + 60% C; 17 min—100% C; 19 min—100% A; 25 min—100% A. The quantification was based on peaks areas at 430 nm in comparison with an external standard (DHI Lab Products).

2.5. Intron Analysis

The secondary structure of the intron was deduced, first by identifying the conserved pairs P3–P7, then by deducing with Mfold the remaining parts, using as a guide a set of sequences and structures previously established. For phylogenetic analysis, class E introns present in nuclear rDNA of closely related algae as well as of distant species, were selected. A set of class C introns has also been selected and used as outgroup. The sequences were aligned according to the secondary structure considering only the conserved P3–P8 core, and the tree was constructed with Maximum Likelihood (ML, GTR + Γ; 1000 bootstraps), as previously described [20,21].

2.6. 18S rDNA Phylogeny

For identification, sequence of 18S rDNA excluding the introns and complete ITS1-5.8S-ITS2 region of isolate *nl3* was searched against non-redundant nucleotide database for homologous sequences using online blast program (BLASTN) (http://www.ncbi.nlm.nih.gov/BLAST/). Clustal X2.1 software was then employed for the automatic multiple alignments of homologous sequences and studied sequence. A preliminary analysis with Paup was carried out [22]. Phylogenetic trees were generated using maximum likelihood (ML, GTR, G + I:4 model) with TREEFINDER [23], distance (neighbour-joining, K2 model) and maximum parsimony (MP) with MEGA7 [24]), with 1000 bootstraps.

3. Results

3.1. Sequencing of the rDNA-ITS Region

A PCR fragment was obtained using the primers NS1 and ITS4 (Table 1) for both strains named *nl3* and *nl6*. These primers, originally described to amplify the 18S-ITS1-5.8S-ITS2 rDNA region in fungi [25], proved also to work well for land plants [26]. The PCR fragment has a size of around 3500 bp for *nl3* and 2600 bp for *nl6*. PCR products were sequenced using the primers listed in Table 1 in both forward and reverse orientations and blasted against the sequences deposited on NCBI database. The BLAST search results showed for *nl6* a 100% identity with *N. salina* (D12, accession number JX185299.1), while *nl3* proved to be closer although not identical to *Desmodesmus* sp. GM4i (AB917136.1). The sequence of *Desmodesmus* sp. *nl3* is deposited in GenBank (MN746324). *Desmodesmus* sp. *nl3* and *Desmodesmus* sp. GM4i, both have an intron inserted at the same position in the 18S rDNA, S516 (referring to the insertion in the rDNA of *E. coli*), although the *nl3* is longer in size (754 pb versus 404 bp). The presence of S516 intron in *Desmodesmus* GM4i has already been reported [27]. Therefore, we analyzed the secondary structure of the *nl3* S516 intron. It comprises the 9 typical stem-loop structures of group I introns, with an additional loop of 347 bp at the level of the P9.3 branch (Figure 1). Overall, the deduced structure and sequence variations strongly indicate that the intron belongs to the E class [28], probably of the E2 type. Another group I intron, S1046, class C, is also present in the 18S rDNA of the strain *nl3*.

Table 1. Primers used in this study.

Primer	Sequence (5'-3')	DNA Region Amplified
NS1	GTAGTCATATGCTTGTCTC	SSU
NS2	GGCTGCTGGCACCAGACTTGC	SSU
NS3	GCAAGTCTGGTGCCAGCAGCC	SSU
NS4	CTTCCGTCAATTCCTTTAAG	SSU
NS5	AACTTAAAGGAATTGACGGAAG	SSU
NS8	TCCGCAGGTTCACCTACGGA	SSU
ITS1	TCCGTAGGTGAACCTGCGG	ITS1
ITS3	GCATCGATGAAGAACGCAGC	ITS2
ITS4	TCCTCCGCTTATTGATATGC	ITS2

3.2. Phylogenetic Analysis of the nl3 Sequence

In order to position *nl3* at the molecular level, phylogenetic analyses were performed using a set of selected microalgal rDNAs including some closely related sequences where introns present were excluded. The results show that isolate *nl3* branches with two strains of *Desmodesmus subspicatus* (Figure 2). Therefore, isolate *nl3* is a green microalgal species, belonging to the genus *Desmodesmus*, family Scenedesmaceae.

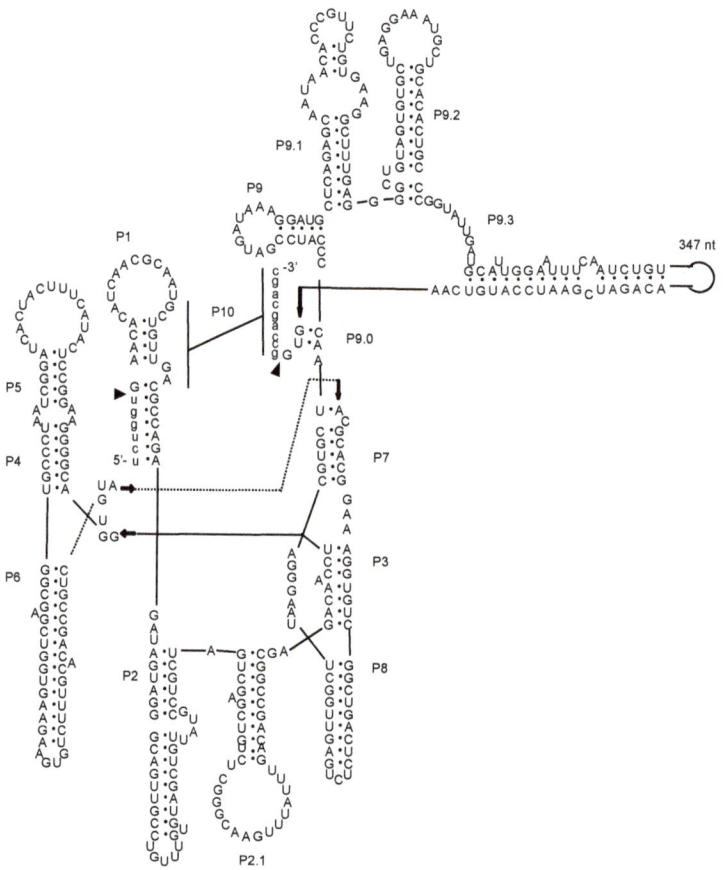

Figure 1. Predicted secondary structure of the group I intron S516 of *Desmodesmus* sp. *nl3*, located in the 18S rDNA. Paired elements are indicated as P1 to P9, the intron sequence is in upper case letters, the exon sequence in lower case letters.

3.3. Microalgal Growth

The morphology of both isolates was observed under the microscope. Cells of *Desmodesmus* sp. *nl3* are oval–shaped with a diameter ranging between 6 and 10 µM while cells of *N. salina nl6* are smaller, with a diameter ranging between 3 and 5 µM. F/2 medium was utilized for the investigation of the growth of *N. salina nl6* (Figure 3a–c). A salinity of 24‰ representing the average salinity found in the shrimp ponds where the microalgal sample has been collected was chosen. The temperature was 25 °C and the light intensity was 200 µmol m^{-2}s^{-1}. The specific growth rate of *nl6* was 0.13 day^{-1} and the biomass yield was 0.39 ± 0.01 g L^{-1} at the entry of the stationary phase. The specific growth rate of *nl6* was 0.13 day^{-1} and the biomass yield was 0.39 ± 0.01 g L^{-1} at the entry of the stationary phase. The specific growth rate was consistent with that reported for other *Nannochloropsis* isolates (from 0.073 to 0.21 day^{-1}) and was lower than that of *N. salina* CCMP537 and CCMP1176 (0.19 day^{-1}) [29], probably because 2% (*v/v*) CO_2 was supplemented in the F/2 medium for these last two strains. We noticed a change in color during cultivation, from green in lag and exponential phases to yellowish in stationary phase. This phenomenon was also reported by [30,31]. This color change is even more obvious when cells are cultivated on agar plates (Figure 3b,c) where the cells turn orange.

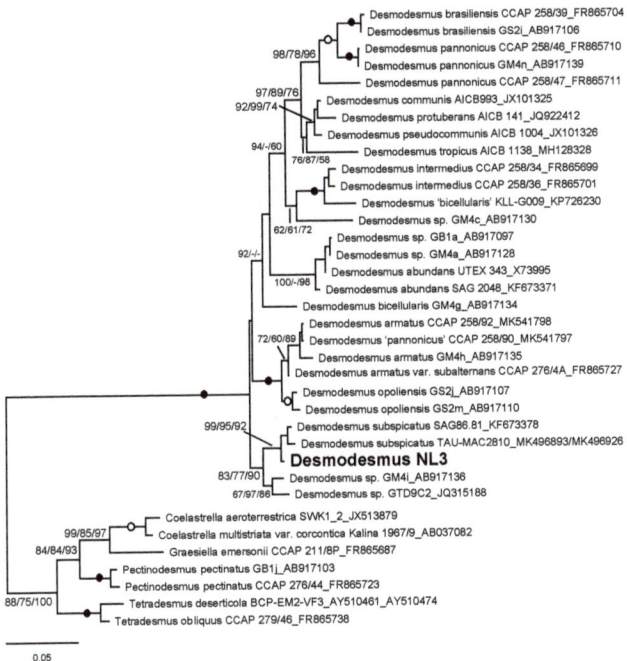

Figure 2. Phylogenetic position of isolate nl3 within Scenedesmaceae. The phylogenetic tree was built on a selected set of microalgal rDNA (18S–ITS1-5.8S-ITS2) sequences available in GenBank, using maximum likelihood (ML), neighbor-joining (NJ) and maximum parsimony (MP). Bootstrap values after 1000 replicates are indicated at nodes for mL/NJ/MP. Filled and open circles indicate 100 or >95% BV support with all methods, respectively; hyphen, node not supported.

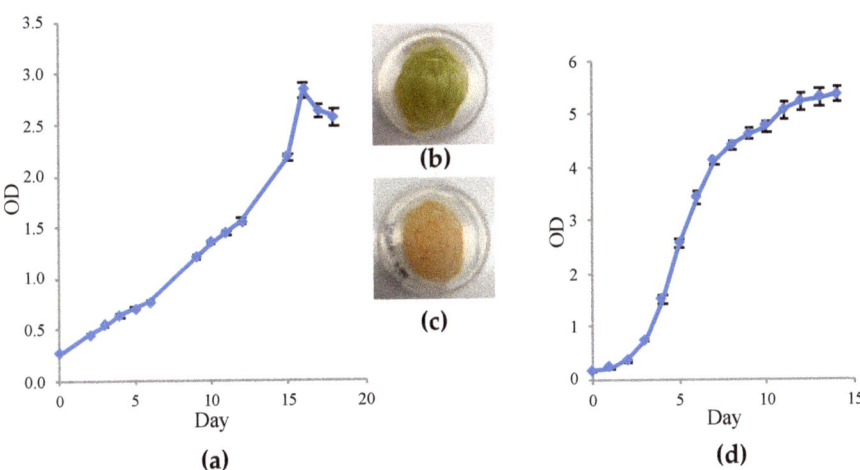

Figure 3. Growth curves of the two microalgal isolates. (**a**) growth curve of *N. salina nl6* in liquid medium (n = 3, 24‰ NaCl, F/2 medium, 200 µmol m^{-2}s^{-1}); (**b**) Agar plate of *nl6* after 10 days growth; (**c**) Agar plate of *nl6* after several weeks of growth; (**d**) growth curve of *Desmodesmus* sp. *nl3* in liquid medium (n = 3, TAP medium, 200 µmol m^{-2}s^{-1}).

For *nl3*, although the initial isolation was made on F/2 medium (see material and methods), growth was analyzed on Tris-Acetate-Phosphate medium (TAP), which is routinely used for the growth of *Chlamydomonas* [32], a freshwater green microalga belonging to the same class (Chlorophyceae) because *nl3* easily grows on this medium. As shown in Figure 3d, *nl3* had a two–day lag phase before entering the exponential phase from day 3 to day 12. The specific growth rate of *nl3* was 0.27 day^{-1}. The cells enter the stationary phase at day 14 with biomass yield of 1.54 ± 0.06 g L^{-1}. The biomass yield was higher than those in other studies. For instance, Ji et al., (2013) [33] reported a maximum biomass of 0.758 g L^{-1} from *Desmodesmus* sp. EJ15-2 cultivated in optimal condition of 30 °C, 98 μmol m^{-2} s^{-1}, BG11 medium and 14:10 (Light:Dark) for 14 days. In another study, Zhang et al., (2016) reported biomass yields ranging from 0.520 to 0.792 g L^{-1} in seven different *Desmodesmus* sp. strains cultivated in BG11 and maintained for 19 days at an irradiance of 80 μmol m^{-2} s^{-1} and 14:10 (Light:Dark) at 25 ± 1 °C [34]. These differences are probably due to the presence of acetate in our medium which provides an additional carbon source for growth, besides atmospheric CO_2.

3.4. Pigment Content in the Exponential and Stationary Phases

We then evaluated the pigment content of both strains in mid-exponential and early stationary phase (Table 2). As expected, chlorophyll b is lacking in *N. salina nl6*, as the *Nannochloropsis* genus lacks this type of chlorophyll [35] while it is present in *Desmodesmus* sp. *nl3* which belongs to Chlorophyta. In both strains, most of the pigments decrease upon stationary phase, especially chlorophyll a as already observed in *N. oceanica* [36] and other microalgae [37]. Solovchenlo et al., (2014) [36] propose that this is an indication of a down-sizing and a remodeling of the light harvesting antenna of the microalga following nutrient stress. On the contrary, astaxanthin (0.36 ± 0.03 mg/g DW) and canthaxanthin (0.18 ± 0.01 mg/g DW), two pigments of potential for commercial success [8,38] are detected in the stationary phase of *N. salina nl6*. The content of astaxanthin measured is comparable to that reported in [39] when expressed to the amount of chlorophyll a (2.9% in our case and 2.6% in their case). Associated with the decrease of chlorophyll a, the presence of astaxanthin explains the color change occurring between the exponential and the stationary phase in *N. salina nl6*. When pigment analysis is made on agar plates maintained for several weeks in the light as depicted in Figure 3c, higher amounts of astaxanthin (2.52 ± 1.3 mg/g DW) are present, although far less than the amounts obtained for the green alga *Haematoccocus pluvialis* (8–36 mg/g DW) [40], the well-known astaxanthin producer. Nevertheless, these latter results suggest that astaxanthin production could be optimized in *N. salina nl6*.

On the other hand, lutein is detected in *Desmodesmus* sp. *nl3* and the amount does not vary significantly between exponential (1.88 ± 0.11 mg DW) and stationary (1.63 ± 0.24 mg DW) phase (Table 2). This amount is in the range of that found in *Scenedesmus obliquus*, a green microalga of the same family [41]. In addition, as it is shown that the lutein content may vary upon various stresses [42], we cultivated *Desmodesmus* sp. *nl3* under various salinities (between 10 and 35‰) (Table 2). We noticed a significant increase of the lutein content upon salt addition, with the 20‰ salinity showing the highest lutein content (7.00 ± 0.24 mg/g) at the beginning of the stationary phase. The biomass yield does not change much in the different salinity conditions (Table 2), meaning that the lutein content reaches 9.87 mg L^{-1} in 20‰ salinity against 2.51 mg L^{-1} when cells are cultivated in fresh water medium. The photoprotective mechanisms in the presence of salt and light which are activated to avoid the presence of reactive oxygen species will be worthy to study, knowing that differences exist inside Chlorophyta [43]. The protein content was determined in the different salinities tested and shown to represent around 40% of the DW (Table 2).

Table 2. Pigment content in exponential and stationary phase (mg/g DW). *N. salina* nl6 was cultivated in F/2 medium (24‰ salinity) and *Desmodesmus* sp. nl3 on TAP medium; Lutein, protein, and biomass contents of *Desmodesmus* sp. nl3 (n = 3).

Pigment	N. salina nl6		Desmodesmus sp. nl3	
	Exponential Phase	Stationary Phase	Exponential Phase	Stationary Phase
Lutein	0.48 ± 0.02	0.22 ± 0.03	1.88 ± 0.11	1.63 ± 0.24
Neoxanthin	0.13 ± 0.00	nd	0.16 ± 0.02	0.13 ± 0.08
Violaxanthin	2.51 ± 0.11	0.59 ± 0.10	0.61 ± 0.05	0.31 ± 0.07
Antheraxanthin	nd	nd	nd	0.13 ± 0.03
Chlorophyll b	-	-	3.79 ± 0.45	2.16 ± 0.09
Chlorophyll a	12.20 ±0.49	3.07 ± 0.36	14.16 ± 0.64	6.82 ± 0.33
Beta-carotene	0.27 ± 0.00	0.06 ± 0.01	0.52 ± 0.05	0.18 ± 0.02
Astaxanthin	nd	0.36 ± 0.03	nd	nd
Canthaxanthin	nd	0.18 ± 0.01	0.02 ± 0.00	0.01 ± 0.00

Desmodesmus sp. nl3 (Stationary Phase)	Salinity				
	0‰	10‰	20‰	30‰	35‰
Lutein	1.63 ± 0.24	4.01 ± 0.55	7.00 ± 0.24	5.67 ± 0.17	2.24 ± 0.72
Protein (% DW)	26.3 ± 1.99	40.54 ± 2.96	38.07 ± 0.29	39.26 ± 1.71	31.93 ± 2.55
Biomass yield (g L^{-1})	1.54 ± 0.06	1.19 ± 0.04	1.41 ± 0.03	1.21 ± 0.02	1.12 ± 0.00

3.5. Fatty Acid Profile and Content in the Exponential and Stationary Phases

We then examined the fatty acid profile and content in exponential and onset of the stationary phase in *N. salina* nl6 and *Desmodesmus* sp. nl3 in the growth conditions mentioned above (F/2 medium and 24‰ salinity for *N. salina* nl6 and TAP medium for *Desmodesmus* sp. nl3. As indicated in Table 3 (left), *N. salina* nl6 fatty acid content is 18.87 ± 3.21% of DW in exponential phase and reaches 40.41 ± 2.87 of DW in stationary phase, indicating that this strain is a bona fide oleaginous microalga. The fatty acid profile is typical of *Nannochloropsis* species, and characterized by four main fatty acids: palmitic (16:0), palmitoleic (16:1n−7), arachidonic (20:4n−6) and eicosapentaenoic (20:5n−3) (EPA) acids [44].

Table 3. Fatty acid profiles of *N. salina* nl6 (F/2 medium 24‰ salinity) and *Desmodesmus* sp. nl3 (TAP medium) in exponential and stationary phases (n = 3).

Fatty Acid	N. salina nl6		Desmodesmus sp. nl3	
	Exponential Phase	Stationary Phase	Exponential Phase	Stationary Phase
C_{14}	nd	0.68 ± 0.21	nd	nd
C_{16}	55.75 ± 1.86	44.62 ± 1.39	24.81 ± 0.62	25.70 ± 1.75
$C_{16:1}$ (C16:1n−7)	35.64 ± 2.15	33.61 ± 1.24	nd	nd
C_{18}	nd	1.28 ± 0.56	1.60 ± 0.81	nd
$C_{18:1-cis}$	2.79 ± 0.86	7.07 ± 0.78	12.12 ± 0.95	27.73 ± 1.27
$C_{18:2-cis}$	nd	0.65 ± 0.24	11.20 ± 0.81	21.74 ± 0.41
$C_{18:3(cis-\Delta 9,12,15)}$	-	-	50.27 ± 0.32	24.83 ± 2.20
$C_{20:3}$ (C20:3n−6)	1.01 ± 0.42	1.11 ± 0.15	-	-
$C_{20:4}$ (C20:4n−6)	0.42 ± 0.28	4.01 ± 0.29	-	-
$C_{20:5}$ (C20:5n−3)	4.39 ± 0.73	7.08 ± 0.39	-	-
Σ SFA [a]	55.75 ± 1.86	46.47 ± 1.20	26.41 ± 1.43	25.70 ± 1.75
Σ MUFA [b]	38.43 ± 1.60	40.68 ± 1.00	12.12 ± 0.95	27.73 ± 1.27
Σ PUFA [c]	5.82 ± 1.33	12.86 ± 0.95	61.47 ± 0.48	46.57 ± 2.44
% DW	18.87 ± 3.21	40.41 ± 2.87	10.22 ± 0.91	8.65 ± 0.50

[a] SFA: saturated fatty acids; [b] MUFA: monounsaturated fatty acids, [c] PUFA: polyunsaturated fatty acids.

The total fatty acid content of *Desmodesmus* sp. nl3 is around 10% DW (Table 3 right), which is expected since green microalgae are not considered as oleaginous. The fatty acid profile is typical of green microalgae, with C16:0 and C18:3 as main fatty acids [45]. The chain lengths are between C_{14} and C_{18}, as commonly found in members of Chlorophyta [46].

We then investigated whether these strains would present similar fatty acid profiles in other growth conditions (Tables 4 and 5). For that purpose, we cultivated microalgal cells at different salinities, 10‰, 20‰, 30‰, and 35‰, and harvested at the onset of the stationary phase. Across the four salinity conditions, fatty acid profiles of both strains varied slightly. The biomass yields do not vary much either (see Table 4 for *N. salina nl6* and Table 2 for *Desmodesmus* sp. *nl3*).

Table 4. Fatty acid profiles of *N. salina nl6* in different salinities (n = 2) harvested in stationary phase.

N. salina nl6	Salinity			
Fatty Acid	10‰	20‰	30‰	35‰
C_{14}	2.81 ± 0.03	2.90 ± 0.01	3.89 ± 0.09	1.94 ± 0.51
C_{16}	48.50 ± 5.25	43.85 ± 2.14	43.86 ± 0.59	44.41 ± 1.35
$C_{16:1}$ (C16:1n−7)	40.06 ± 3.99	38.25 ± 1.01	36.37 ± 0.82	37.59 ± 1.58
C_{18}	nd	1.52 ± 0.83	1.24 ± 0.36	0.63 ± 0.16
$C_{18:1-cis}$	8.63 ± 1.28	11.25 ± 0.63	10.88 ± 1.10	10.89 ± 0.29
$C_{18:2-cis}$	nd	0.40 ± 0.04	0.57 ± 0.09	0.18 ± 0.25
C_{20}	nd	nd	nd	nd
$C_{20:3}$ (C20:3n−6)	nd	nd	nd	0.01 ± 0.00
$C_{20:4}$ (C20:4n−6)	nd	0.97 ± 0.09	1.51 ± 0.38	1.93 ± 0.24
$C_{20:5}$ (C20:5n−3)	nd	0.86 ± 0.19	1.68 ± 0.47	2.44 ± 0.31
Σ SFA [a]	51.31 ± 5.28	48.27 ± 1.31	48.99 ± 1.04	46.98 ± 1.00
Σ MUFA [b]	48.69 ± 5.28	49.50 ± 1.65	47.26 ± 0.28	48.47 ± 1.30
Σ PUFA [c]	nd	2.23 ± 0.33	3.76 ± 0.76	4.55 ± 0.29
Biomass yield (g/L)	0.4 ± 0.01	0.44 ± 0.00	0.49 ± 0.00	0.48 ± 0.03

[a] SFA: saturated fatty acids; [b] MUFA: monounsaturated fatty acids, [c] PUFA: polyunsaturated fatty acids.

Table 5. Fatty acid profiles of *Desmodesmus* sp. *nl3* in different salinities (n = 3) harvested in stationary phase.

Desmodesmus sp. *nl3*	Salinity			
% Fatty Acids	10‰	20‰	30‰	35‰
C_{14}	nd	nd	nd	nd
C_{16}	19.00 ± 0.33	17.54 ± 1.95	18.05 ± 0.63	21.53 ± 0.51
$C_{16:1}$	nd	nd	1.51 ± 0.27	3.62 ± 0.36
C_{18}	nd	nd	nd	0.55 ± 0.54
$C_{18:1-cis}$	9.47 ± 0.97	6.04 ± 0.21	14.64 ± 0.51	26.23 ± 2.69
$C_{18:2-cis}$	27.70 ± 0.89	22.71 ± 1.01	23.60 ± 1.24	19.22 ± 0.47
$C_{18:3(cis-\Delta9,12,15)}$	43.83 ± 1.79	53.71 ± 2.35	42.20 ± 0.78	29.03 ± 3.79
Σ SFA[a]	19.00 ± 0.33	17.54 ± 1.95	18.05 ± 0.63	21.90 ± 0.93
Σ MUFA[b]	9.47 ± 0.97	6.04 ± 0.21	16.14 ± 0.43	29.85 ± 2.85
Σ PUFA[c]	71.53 ± 1.30	76.42 ± 1.76	65.80 ± 0.58	48.26 ± 3.78

[a] SFA: saturated fatty acids; [b] MUFA: monounsaturated fatty acids, [c] PUFA: polyunsaturated fatty acids.

3.6. Effect of Salinity Conditions on Biodiesel Quality

The potential of the microalgal oils of both strains for biodiesel production was investigated by calculating seven biodiesel properties: average degree of unsaturation (ADU), kinematic viscosity (kV, 40 °C, mm^2 s^{-1}), specific gravity (SG, kg L^{-1}), cloud point (CP, °C), cetane number (CN), iodine value (IV, g I$_2$ 100g^{-1}), and higher heating value (HHV, (Mj kg^{-1})) (see Materials and Methods) based on the fatty acid contents and profiles (Tables 6 and 7). Those properties were evaluated through the comparison with the US (ASTMD 6751-08) and Europe (EN 14214) biodiesel standards. As depicted in these tables, all the six values agreed with the requirements of the two standards for *N. salina nl6* while iodine parameter (IV) did not for *Desmodesmus* sp. *nl3*.

Table 6. Biodiesel properties of *N. salina nl6* in different salinities and growth phases.

Biodiesel Properties	Salinity (Stationary Phase)				Stationary Phase	Exponential Phase	Standards	
	10‰	20‰	30‰	35‰	24‰	24‰	US (ASTM D 6751-08)	Europe (EN 14214)
ADU	0.49 ± 0.05	0.58 ± 0.00	0.63 ± 0.04	0.69 ± 0.01	0.97 ± 0.04	0.65 ± 0.06	-	-
kV	4.90 ± 0.03	4.84 ± 0.00	4.81 ± 0.03	4.77 ± 0.00	4.6 ± 0.02	4.8 ± 0.04	1.9–6.0	3.5–5.5
SG	0.88 ± 0.00	0.88 ± 0.00	0.88 ± 0.00	0.88 ± 0.00	0.88 ± 0.00	0.88 ± 0.00	0.85–0.9	-
CP (°C)	13.49 ± 0.70	12.18 ± 0.03	11.60 ± 0.53	10.81 ± 0.09	7.07 ± 0.47	11.3 ± 0.74	-	-
CN	59.63 ± 0.35	58.98 ± 0.01	58.69 ± 0.26	58.29 ± 0.05	56.42 ± 0.24	58.54 ± 0.37	min 47	min 51
IV	48.93 ± 3.92	56.20 ± 0.15	59.45 ± 2.95	63.83 ± 0.51	84.68 ± 2.64	61.12 ± 4.13	-	Max 120
HHV	39.39 ± 0.09	39.56 ± 0.00	39.64 ± 0.07	39.74 ± 0.01	40.24 ± 0.06	39.68 ± 0.10	-	-

Table 7. Biodiesel properties of *Desmodesmus sp. nl3* in different salinities and growth phases.

Biodiesel Properties	Salinity (Stationary Phase)				Stationary Phase	Exponential Phase	Standards	
	10‰	20‰	30‰	35‰	0‰	0‰	US (ASTM D 6751-08)	Europe (EN 14214)
ADU	1.96 ± 0.03	2.13 ± 0.06	1.90 ± 0.01	1.55 ± 0.08	1.46 ± 0.06	1.85 ± 0.02	-	-
kV	3.97 ± 0.02	3.86 ± 0.04	4.01 ± 0.01	4.23 ± 0.05	4.29 ± 0.04	4.04 ± 0.01	1.9–6.0	3.5–5.5
SG	0.88 ± 0.00	0.88 ± 0.00	0.88 ± 0.00	0.88 ± 0.00	0.88 ± 0.00	0.88 ± 0.0	0.85–0.9	-
CP (°C)	−6.23 ± 0.44	−8.40 ± 0.79	−5.38 ± 0.15	−0.76 ± 1.13	0.54 ± 0.83	−4.76 ± 0.21	-	-
CN	49.78 ± 0.22	48.70 ± 0.39	50.21 ± 0.08	52.51 ± 0.57	53.16 ± 0.42	50.52 ± 0.11	min 47	min 51
IV	158.74 ± 2.47	170.81 ± 4.40	153.97 ± 0.86	128.27 ± 6.31	121.06 ± 4.65	150.53 ± 1.19	-	Max 120
HHV	41.99 ± 0.06	42.28 ± 0.10	41.88 ± 0.02	41.27 ± 0.15	41.10 ± 0.11	41.80 ± 0.03	-	-

4. Discussion

In this study, two microalgal isolates from the Ninh Thuan province of Vietnam have been genetically identified and characterized in terms of growth at lab-scale and biomass composition.

The *nl6* isolate is identical to *N. salina* D12 recorded in GenBank (JX185299.1) based on 18S rDNA –ITS sequence. The strain exhibits a high percentage of fatty acids at the beginning of the stationary phase (40% DW), and a high percentage of saturated fatty acids (i.e., 46% of SFA), suggesting *nl6* as a promising candidate for biodiesel production. In addition, there is also a high proportion of MUFAs (nearly 50%), which is an important index for the evaluation of biodiesel quality [29,47]. Indeed, in biodiesel, oxidative stability is improved by saturated fatty esters due to their high cetane number while low-temperature properties are favored by unsaturated fatty esters [48]. Amongst these MUFAs, besides its role in biodiesel, palmitoleic acid (C16:1), which represents 30–40% of the total fatty acids has generated attention in recent years due to its nutritional value [49], and its use in the production of linear low-density polyethylene [50]. Concerning PUFAS, and as quite well known for long time, EPA is also interesting for human health and represents between 4 and 7% of the total fatty acids, which is in the range observed for this genus [51]. Pigment investigations also outlined a substantial increase of astaxanthin and canthaxanthin in stationary phase. Therefore, this microalga is attractive both for biodiesel, and its nutritional value, which makes it appreciated in aquaculture. The fact that the fatty acid profile does not change much in a large range of salinity (10–35‰) makes this isolate particularly attractive for cultivation in open ponds close to the sea in tropical climates where heavy rains during monsoon are responsible for fluctuating salinities. The disadvantage we observe here concerns its low specific growth rate which should be improved to make the strain economically competitive.

The *nl3* isolate belongs to the *Desmodesmus* genus and phylogenetic analysis based on the 18S rDNA–ITS sequence shows that the strain groups with *D. subspicatus*. The strain has two group I introns, one S516 and S1046. The S516 group I intron was shown to display a typical secondary structure, which can be assigned to class E. Group I introns are found in organellar genomes of almost all eukaryotes and nuclear rDNA genes of unicellular eukaryotes such as fungi, algae and amoebae [20,52] and are able to self-splice at least in vitro to give the mature RNA sequence. Some of them also have a homing endonuclease (HE) gene that increases their ability to spread [52]. Nearly all the S516 introns of the Scenedesmaceae family analyzed in our study including that of *Desmodesmus* sp. *nl3* are grouped in the phylogenetic tree based on the conserved P3–P8 structure of the introns (Figure S1a). When considering the entire sequence of the introns excluding endonuclease gene, pairwise similarity analysis suggests that the *nl3* intron is much closer to the introns of *D. intermedius* (71.5% similarity) and *D. opoliensis* GS2m and GS2j (64–67% similarity, respectively), than to that of *Coelastrella* (46.5% similarity) (Figure S1b). In addition, the group I intron of *D. opoliensis* GS2j is longer than all others and has an HE with a His-Cys box domain [36] (Figure S2), suggesting that this intron could be mobile. HEs typically recognize sequences of 18–24 bp in intron-less sequences and introduce a double strand break. The double strand break is then repaired by copying the intron sequence in the intron-less DNA sequence [52]. Interestingly, the HE in the *D. opoliensis* GS2j intron is inserted into P9.3, which is also the branch whose length varies between the analyzed introns. Available data suggest that the introns in the algae studied here may have a common origin, subsequently either persisting (*D. opoliensis* GS2j) or changing with the degradation (*D. intermedius*, *D. opoliensis* GS2m, *Desmodesmus* sp. *nl3*) and final loss of HE (*Desmodesmus* GM4i), and the intron itself, consistent with the homing life cycle of invasion and loss of the introns in closely related species [53].

The biodiesel properties of the fatty acids of *Desmodesmus* sp. *nl3* do not match with the US and European standards because of the iodine value. However, this strain has other interesting characteristics, such as its high lutein content, which, coupled to its high protein content, makes it attractive for the fish industry since lutein supplementation (50 mg kg^{-1}) improves survival and promotes efficient carotenoid pigmentation of the skin of goldfish juveniles [54]. In addition, C18:3 (α-linolenic acid), represents between 40 and 50% of the total fatty acid content in nearly all the tested conditions. This fatty acid is usually in low amounts in vegetal oils and has also been detected in substantial amounts in other green microalgal isolates from Vietnam [55]. Like for *N. salina nl6*, the tolerance of *nl3* to various salinities also makes it interesting for cultivation in coastal regions located in tropical climates. In addition, since this strain is able to assimilate organic carbon source like acetate and ammonium as nitrogen source, it is also interesting for combining wastewater remediation and use of the resulting algal biomass for aquaculture and fisheries purposes.

5. Conclusions

Our results show the potential of two microalgal isolates for cultivation in coastal regions of Vietnam. *Desmodesmus* sp. *nl3* possesses a group I intron in the 18S rDNA sequence, closely related to other group I introns of the *Desmodesmus* genus. This robust species reaches decent biomass yield with high protein and lutein contents at the beginning of the stationary in lab-scale conditions, tolerates various salinities and is able to assimilate an organic carbon source and ammonium. These characteristics makes it attractive for wastewater treatment coupled to aquaculture purposes. *N. salina nl6* also tolerates various salinities and has an attractive fatty acids profile and content including PUFAs which could be exploited for biodiesel and nutraceutical aspects. The fact that the two strains with contrasting characteristics have been isolated in the same ponds in the Ninh Thuan province of Vietnam reflects the microalgal diversity found in this region, where growth conditions vary greatly in terms of water availability and salinity depending on the season. For outdoor cultivation, future analyses should focus on the tolerance to other stresses such as light intensity and temperature.

Supplementary Materials: The following are available online at http://www.mdpi.com/1996-1073/13/4/898/s1, Figure S1: Analysis of *Desmodesmus* sp. *nl3* S516 intron. (a) Phylogenetic analysis of S516 group I introns of IE and IC1 classes amongst microalgae based on the P3–P8 conserved structures, (b) pairwise analysis of group I introns of the Scenedesmaceae family. Figure S2: Amino acid sequences of homing endonucleases (HE) of *Naegleria jamiesoni* (Njam, AAB71747.1), *Porphyra umbilicalis* (Pumb, AAV35433), *Allovahlkampfia spelaea* (Aspe, ABD62811), *Coemansia mojavensis* (Cmoj, BAB87243), *Naegleria philippinensis* (Nphil, CAJ44447), *Desmodesmus opoliensis* GS2j (DopS2j, AB917110), *Sclerotinia sclerotiorum* (Sscl, XP_001587714). Color code (according Clustal X color scheme): blue: hydrophobic residues; red: positively charged residues; magenta: negatively charged residues; green: polar residues; cyan: aromatic residues, orange: glycine residue, yellow: proline residue. See [21] for detailed analysis of HE.

Author Contributions: Conceptualization, G.E. and C.R.; Funding acquisition, H.A.L. and G.E.; Methodology, G.E. and C.R.; Supervision, G.E. and C.R.; Validation, T.N.L., Z.A., A.C., D.C., G.E. and C.R.; Writing–original draft, T.N.L.; Writing–review & editing, T.N.L., D.C., H.A.L., G.E. and C.R. All authors have read and agreed to the published version of the manuscript.

Funding: The authors wish to thank the Académie de Recherche et d'Enseignement Supérieur (ARES-CCD, Brussels, Belgium) for their financial support in the frame of the RENEWABLE project (REmoval of NutriEnts in Wastewater treatments via microAlgae and BiofueL/biomass production for Environmental sustainability in Vietnam, PRD 2016-2020). A.C. is the recipient of an Action de Recherche Concertée doctoral fellowship from the University of Liege (DARKMET ARC grant 17/21-08).

Acknowledgments: The authors wish to thank M. Radoux for expert technical assistance.

Conflicts of Interest: The authors declare no conflict of interest.

References

1. Rodolfi, L.; Chini, G.; Niccol, Z.; Giulia, B.; Natascia, P.; Gimena, B.; Mario, B.; Tredici, R. Microalgae for oil: Strain selection, induction of lipid synthesis and outdoor mass cultivation in a low-cost photobioreactor. *Biotechnol. Bioeng.* **2009**, *102*, 100–112. [CrossRef] [PubMed]
2. Sydney, E.B.; Sydney, A.C.N.; de Carvalho, J.C.; Soccol, C.R. Microalgal strain selection for biofuel production. In *Biomass, Biofuels and Biochemicals: Biofuels from Algae*; Panday, A., Chang, J.S., Soccol, C.R., Lee, D.J., Chisti, Y., Elsevier, B.V., Eds.; Elsevier: Amsterdam, The Netherlands, 2019; pp. 59–66.
3. Laurens, L.M.L.; Chen-Glasser, M.; McMillan, J.D. A perspective on renewable bioenergy from photosynthetic algae as feedstock for biofuels and bioproducts. *Algal Res.* **2017**, *24*, 261–264. [CrossRef]
4. Hu, H.; Gao, K. Optimization of growth and fatty acid composition of a unicellular marine picoplankton, Nannochloropsis sp., with enriched carbon sources. *Biotechnol. Lett.* **2003**, *25*, 421–425. [CrossRef] [PubMed]

5. Sajjadi, B.; Chen, W.Y.; Raman, A.A.A.; Ibrahim, S. Microalgae lipid and biomass for biofuel production: A comprehensive review on lipid enhancement strategies and their effects on fatty acid composition. *Renew. Sustain. Energy Rev.* **2018**, *97*, 200–232. [CrossRef]
6. Petrie, J.R.; Singh, S.P. Expanding the docosahexaenoic acid food web for sustainable production: Engineering lower plant pathways into higher plants. *AoB Plants* **2011**, *2011*. [CrossRef]
7. Spolaore, P.; Joannis-Cassan, C.; Duran, E.; Isambert, A. Commercial applications of microalgae. *J. Biosci. Bioeng.* **2006**, *101*, 87–96. [CrossRef]
8. Gong, M.; Bassi, A. Carotenoids from microalgae: A review of recent developments. *Biotechnol. Adv.* **2016**, *34*, 1396–1412. [CrossRef]
9. De Vries, J.; Archibald, J.M. Endosymbiosis: Did Plastids Evolve from a Freshwater Cyanobacterium? *Curr. Biol.* **2017**, *27*, R103–R122. [CrossRef]
10. De Vries, J.; Gould, S.B. The monoplastidic bottleneck in algae and plant evolution. *J. Cell Sci.* **2018**, *131*, jcs203414. [CrossRef]
11. Archibald, J.M. Genomic perspectives on the birth and spread of plastids. *Proc. Natl. Acad. Sci. USA* **2015**, *112*, 10147–10153. [CrossRef]
12. Keeling, P.J. The Number, Speed, and Impact of Plastid Endosymbioses in Eukaryotic Evolution. *Annu. Rev. Plant Biol.* **2013**, *64*, 583–607. [CrossRef] [PubMed]
13. Newman, S.M.; Boynton, J.E.; Gillham, N.W.; Randolph-Anderson, B.L.; Johnson, A.M.; Harris, E.H. Transformation of chloroplast ribosomal RNA genes in Chlamydomonas: Molecular and genetic characterization of integration events. *Genetics* **1990**, *126*, 875–888. [PubMed]
14. Harris, E.H. *The Chlamydomonas Sourcebook*; Elsevier Inc.: Amsterdam, The Netherlands, 1989.
15. Guillard, R.R.; Ryther, J.H. Studies of marine planktonic diatoms. I. Cyclotella nana Hustedt, and Detonula confervacea (cleve) Gran. *Can. J. Microbiol.* **1962**, *8*, 229–239. [CrossRef] [PubMed]
16. Makridis, P.; Vadstein, O. Food size selectivity of Artemia franciscana at three. *J. Plankton Res.* **1999**, *21*, 2191–2201. [CrossRef]
17. Wood, A.M.; Everroad, R.C.; Wingard, L.M. Measuring growth rates in microalgal cultures. In *Algal Culturing Techniques*; Andersen, R.A., Ed.; Elsevier: Amsterdam, The Netherlands, 2005; pp. 269–285.
18. Hoekman, S.K.; Broch, A.; Robbins, C.; Ceniceros, E.; Natarajan, M. Review of biodiesel composition, properties, and specifications. *Renew. Sustain. Energy Rev.* **2012**, *16*, 143–169. [CrossRef]
19. Bradford, M.M. A rapid and sensitive method for the quantitation of microgram quantities of protein utilizing the principle of protein-dye binding. *Anal. Biochem.* **1976**, *72*, 248–254. [CrossRef]
20. Wilgenbusch, J.C.; Swofford, D. Inferring Evolutionary Trees with PAUP*. *Curr. Protoc. Bioinforma.* **2003**. [CrossRef]
21. Jobb, G.; Von Haeseler, A.; Strimmer, K. TREEFINDER: A powerful graphical analysis environment for molecular phylogenetics. *BMC Evol. Biol.* **2004**, *4*, 18. [CrossRef]
22. Kumar, S.; Stecher, G.; Tamura, K. MEGA7: Molecular Evolutionary Genetics Analysis Version 7.0 for Bigger Datasets. *Mol. Biol. Evol.* **2016**, *33*, 1870–1874. [CrossRef]
23. White, T.J.; Bruns, T.; Lee, S.; Taylor, J. *Amplification and Direct Sequencing of Fungal Ribosomal RNA Genes for Phylogenetics: PCR—Protocols and Applications—A Laboratory Manual*; Innis, M.A., Gelfand, D.H., Sninsky, J.J., White, T.J., Eds.; Academic Press Inc.: New York, NY, USA, 1990; ISBN 008088671X.
24. Smolik, M.; Krupa-Małkiewicz, M.; Smolik, B.; Wieczorek, J.; Predygier, K. rDNA variability assessed in PCR reactions of selected accessions of Acer. *Not. Bot. Horti Agrobot. Cluj-Napoca* **2011**, *39*, 260–266. [CrossRef]
25. Hoshina, R. DNA analyses of a private collection of microbial green algae contribute to a better understanding of microbial diversity. *BMC Res. Notes* **2014**, *4*, 792. [CrossRef] [PubMed]
26. Suh, S.O.; Jones, K.G.; Blackwell, M. A Group I intron in the nuclear small subunit rRNA gene of Cryptendoxyla hypophloia, an ascomycetous fungus: Evidence for a new major class of Group I introns. *J. Mol. Evol.* **1999**, *48*, 493–500. [CrossRef] [PubMed]
27. Ma, Y.; Wang, Z.; Yu, C.; Yin, Y.; Zhou, G. Evaluation of the potential of 9 Nannochloropsis strains for biodiesel production. *Bioresour. Technol.* **2014**, *167*, 503–509. [CrossRef] [PubMed]
28. Hu, Q.; Xiang, W.; Dai, S.; Li, T.; Yang, F.; Jia, Q.; Wang, G.; Wu, H. The influence of cultivation period on growth and biodiesel properties of microalga Nannochloropsis gaditana 1049. *Bioresour. Technol.* **2015**, *192*, 157–164. [CrossRef] [PubMed]

29. Wu, Z.; Zhu, Y.; Huang, W.; Zhang, C.; Li, T.; Zhang, Y.; Li, A. Evaluation of flocculation induced by pH increase for harvesting microalgae and reuse of flocculated medium. *Bioresour. Technol.* **2012**, *110*, 496–502. [CrossRef]
30. Harris, E.H. Chlamydomonas as a model organism. *Annu. Rev. Plant Physiol. Plant Mol. Biol.* **2001**, *52*, 363–406. [CrossRef]
31. Ji, F.; Hao, R.; Liu, Y.; Li, G.; Zhou, Y.; Dong, R. Isolation of a novel microalgae strain Desmodesmus sp. and optimization of environmental factors for its biomass production. *Bioresour. Technol.* **2013**, *148*, 249–254. [CrossRef]
32. Zhang, Y.; He, M.; Zou, S.; Fei, C.; Yan, Y.; Zheng, S.; Rajper, A.A.; Wang, C. Breeding of high biomass and lipid producing Desmodesmus sp. by Ethylmethane sulfonate-induced mutation. *Bioresour. Technol.* **2016**, *207*, 268–275. [CrossRef]
33. Vieler, A.; Wu, G.; Tsai, C.H.; Bullard, B.; Cornish, A.J.; Harvey, C.; Reca, I.B.; Thornburg, C.; Achawanantakun, R.; Buehl, C.J.; et al. Genome, Functional Gene Annotation, and Nuclear Transformation of the Heterokont Oleaginous Alga Nannochloropsis oceanica CCMP1779. *PLoS Genet.* **2012**, *8*, e1003064. [CrossRef]
34. Solovchenko, A.; Lukyanov, A.; Solovchenko, O.; Didi-Cohen, S.; Boussiba, S.; Khozin-Goldberg, I. Interactive effects of salinity, high light, and nitrogen starvation on fatty acid and carotenoid profiles in Nannochloropsis oceanica CCALA 804. *Eur. J. Lipid Sci. Technol.* **2014**, *116*, 635–644. [CrossRef]
35. Ruivo, M.; Amorim, A.; Cartaxana, P. Effects of growth phase and irradiance on phytoplankton pigment ratios: Implications for chemotaxonomy in coastal waters. *J. Plankton Res.* **2011**, *33*, 1012–1022. [CrossRef]
36. Hu, J.; Nagarajan, D.; Zhang, Q.; Chang, J.S.; Lee, D.J. Heterotrophic cultivation of microalgae for pigment production: A review. *Biotechnol. Adv.* **2018**, *36*, 54–67. [CrossRef]
37. Lubian, L.M.; Montero, O.; Moreno-Garrido, I.; Huertas, I.E.; Sobrino, C.; Gonzalez-Del Valle, M.; Pares, G. Nannochloropsis (Eustigmatophyceae) as source of commercially valuable pigments. *J. Appl. Phycol.* **2000**, *12*, 249–255. [CrossRef]
38. Liu, J.; Sun, Z.; Gerken, H.; Liu, Z.; Jiang, Y.; Chen, F. Chlorella zofingiensis as an alternative microalgal producer of astaxanthin: Biology and industrial potential. *Mar. Drugs* **2014**, *12*, 3487–3515. [CrossRef]
39. Chen, W.-C.; Hsu, Y.-C.; Chang, J.-S.; Ho, S.-H.; Wang, L.-F.; Wei, Y.-H. Enhancing production of lutein by a mixotrophic cultivation system using microalga Scenedesmus obliquus CWL-1. *Bioresour. Technol.* **2019**, *291*, 121891. [CrossRef] [PubMed]
40. Rauytanapanit, M.; Janchot, K.; Kusolkumbot, P.; Sirisattha, S.; Waditee-Sirisattha, R.; Praneenararat, T. Nutrient deprivation-associated changes in green microalga coelastrum sp. TISTR 9501RE enhanced potent antioxidant carotenoids. *Mar. Drugs* **2019**, *17*, 328. [CrossRef] [PubMed]
41. Christa, G.; Cruz, S.; Jahns, P.; de Vries, J.; Cartaxana, P.; Esteves, A.C.; Serôdio, J.; Gould, S.B. Photoprotection in a monophyletic branch of chlorophyte algae is independent of energy-dependent quenching (qE). *New Phytol.* **2017**, *214*, 1132–1144. [CrossRef] [PubMed]
42. Khozin-Goldberg, I.; Boussiba, S. Concerns over the reporting of inconsistent data on fatty acid composition for microalgae of the genus Nannochloropsis (Eustigmatophyceae). *J. Appl. Phycol.* **2011**, *23*, 933–934. [CrossRef]
43. Plancke, C.; Vigeolas, H.; Höhner, R.; Roberty, S.; Emonds-Alt, B.; Larosa, V.; Willamme, R.; Duby, F.; Onga Dhali, D.; Thonart, P.; et al. Lack of isocitrate lyase in Chlamydomonas leads to changes in carbon metabolism and in the response to oxidative stress under mixotrophic growth. *Plant J.* **2014**, *77*, 404–417. [CrossRef]
44. Lang, I.; Hodac, L.; Friedl, T.; Feussner, I. Fatty acid profiles and their distribution patterns in microalgae: A comprehensive analysis of more than 2000 strains from the SAG culture collection. *BMC Plant Biol.* **2011**, *11*, 124. [CrossRef]
45. Knothe, G. Improving biodiesel fuel properties by modifying fatty ester composition. *Energy Environ. Sci.* **2009**, *2*, 759–766. [CrossRef]
46. Ma, X.N.; Chen, T.P.; Yang, B.; Liu, J.; Chen, F. Lipid production from Nannochloropsis. *Mar. Drugs* **2016**, *14*, 61. [CrossRef] [PubMed]
47. Zhou, W.; Wang, H.; Chen, L.; Cheng, W.; Liu, T. Heterotrophy of filamentous oleaginous microalgae Tribonema minus for potential production of lipid and palmitoleic acid. *Bioresour. Technol.* **2017**, *239*, 250–257. [CrossRef]

48. Wang, H.; Gao, L.; Zhou, W.; Liu, T. Growth and palmitoleic acid accumulation of filamentous oleaginous microalgae Tribonema minus at varying temperatures and light regimes. *Bioprocess Biosyst. Eng.* **2016**, *39*, 1589–1595. [CrossRef]
49. Janssen, J.H.; Wijffels, R.H.; Barbosa, M.J. Lipid Production in Nannochloropsis gaditana during Nitrogen Starvation. *Biology* **2019**, *8*, 5. [CrossRef]
50. Hedberg, A.; Johansen, S.D. Nuclear group i introns in self-splicing and beyond. *Mob. DNA* **2013**, *4*, 17. [CrossRef]
51. Corsaro, D.; Köhsler, M.; Venditti, D.; Rott, M.B.; Walochnik, J. Recovery of an Acanthamoeba strain with two group I introns in the nuclear 18S rRNA gene. *Eur. J. Protistol.* **2019**, *68*, 88–98. [CrossRef]
52. Goddard, M.R.; Burt, A. Recurrent invasion and extinction of a selfish gene. *Proc. Natl. Acad. Sci. USA* **1999**, *96*, 13880–13885. [CrossRef]
53. Besen, K.P.; Melim, E.W.H.; da Cunha, L.; Favaretto, E.D.; Moreira, M.; Fabregat, T.E.H.P. Lutein as a natural carotenoid source: Effect on growth, survival and skin pigmentation of goldfish juveniles (Carassius auratus). *Aquac. Res.* **2019**, *50*, 2200–2206. [CrossRef]
54. Thao, T.Y.; Linh, D.T.N.; Si, V.C.; Carter, T.W.; Hill, R.T. Isolation and selection of microalgal strains from natural water sources in Viet Nam with potential for edible oil production. *Mar. Drugs* **2017**, *15*, 194. [CrossRef]
55. Corsaro, D.; Venditti, D. Nuclear Group I introns with homing endonuclease genes in Acanthamoeba genotype T4. *Eur. J. Protistol.* **2018**, *66*, 26–35. [CrossRef] [PubMed]

© 2020 by the authors. Licensee MDPI, Basel, Switzerland. This article is an open access article distributed under the terms and conditions of the Creative Commons Attribution (CC BY) license (http://creativecommons.org/licenses/by/4.0/).

Article

Hydrothermal Liquefaction of Rice Straw Using Methanol as Co-Solvent

Attada Yerrayya [1], A. K. Shree Vishnu [2], S. Shreyas [2], S. R. Chakravarthy [2,3] and Ravikrishnan Vinu [1,3,*]

[1] Department of Chemical Engineering, Indian Institute of Technology Madras, Chennai 600036, India; yerrayya503.nitw@gmail.com
[2] Department of Aerospace Engineering, Indian Institute of Technology Madras, Chennai 600036, India; vishnu.ak@hotmail.com (A.K.S.V.); sshreyas.8@gmail.com (S.S.); src@ae.iitm.ac.in (S.R.C.)
[3] National Centre for Combustion Research and Development, Indian Institute of Technology Madras, Chennai 600036, India
* Correspondence: vinu@iitm.ac.in; Tel.: +91-44-2257-4187

Received: 2 April 2020; Accepted: 28 April 2020; Published: 21 May 2020

Abstract: Hydrothermal liquefaction (HTL) is a promising thermochemical process to treat wet feedstocks and convert them to chemicals and fuels. In this study, the effects of final temperature (300, 325, and 350 °C), reaction time (30 and 60 min), rice-straw-to-water ratio (1:1, 1:5, 1:10, and 1:15 (wt./wt.)), methanol-to-water ratio (0:100, 25:75, 50:50, and 75:25 (vol.%/vol.%)), and alkali catalysts (KOH, NaOH, and K_2CO_3) on product yields, composition of bio-crude, higher heating value (HHV) of bio-crude and bio-char, and energy recovery on HTL of rice straw are investigated. At the optimal processing condition corresponding to the final temperature of 300 °C, 60 min reaction time, and rice-straw-to-water ratio of 1:10 at a final pressure of 18 MPa, the bio-crude yield was 12.3 wt.% with low oxygen content (14.2 wt.%), high HHV (35.3 MJ/kg), and good energy recovery (36%). The addition of methanol as co-solvent to water at 50:50 vol.%/vol.% improved the yield of bio-crude up to 36.8 wt.%. The selectivity to phenolic compounds was high (49%–58%) when only water was used as the solvent, while the addition of methanol reduced the selectivity to phenolics (13%–22%), and improved the selectivity to methyl esters (51%–73%), possibly due to esterification reactions. The addition of KOH further improved the yield of bio-crude to 40 wt.% in an equal composition of methanol:water at the optimal condition. The energy-consumption ratio was less than unity for the methanol and catalyst system, suggesting that the process is energetically feasible in the presence of a co-solvent.

Keywords: hydrothermal liquefaction; rice straw; bio-crude; methanol; phenols; esters; energy-consumption ratio

1. Introduction

Rice straw is one of the abundant lignocellulosic agro residues in the world. India is the second largest producer of rice in the world, with 106 million tons per year, and an annual production of roughly 160 million tons of straw [1]. Owing to mechanized farming, open field burning of rice straw and its stubble has become a common scene in India, especially in the state of Punjab, and this has led to severe greenhouse gas emissions and pollution in the major cities [2]. In Asian countries, rice straw is often burnt in the open fields, leading to airborne emissions that are hazardous to living organisms and the environment. This also leads to the killing of small animals in the fields and a decline of biodiversity [1]. Hence, there is increased attention toward conversion of rice straw to valuable products, such as chemicals, fuels, energy, and bio-products.

Thermochemical techniques, such as pyrolysis, gasification, combustion, and hydrothermal liquefaction (HTL), are used to convert biomass into valuable chemicals, liquid fuels and oils, and gaseous fuel (syngas) [3–5]. Among these, HTL is the only processing technique that can handle wet biomass feedstocks with as high as 50 wt.% moisture, to produce high-quality bio-crude. HTL is generally carried out at 200–350 °C and between 5 and 30 MPa [5,6]. High-pressure operation ensures that water from the biomass does not get converted into a gaseous phase, thus saving the energy required for enthalpy of vaporization. HTL is an accelerated version of geological formation of fossil fuels [5]. While the formation of fossil fuels underneath the earth's surface takes many years, HTL yields liquid crude in a minutes-to-hours timeframe [5]. Unlike water at ambient conditions, the physicochemical properties of water at near-critical and supercritical conditions make it an excellent solvent. Owing to the fewer and weaker hydrogen bonds in hot compressed water than in water at ambient conditions, better solubility of organic molecules is achieved due to the low dielectric constant of the medium. Moreover, the high ionic product of water near the critical temperature aids in acid/base-catalyzed ionic reactions [6,7]. HTL bio-crude is superior to pyrolysis bio-oil, especially in terms of low oxygen content, high carbon content, high calorific value, and low moisture content [6].

HTL of a variety of feedstocks, such as pine wood [8,9], corn stalk [10], lignin [11,12], sewage sludge [13], corn stover [14], switch grass [15], *Nannochloropsis* sp. [16,17], and spirulina [17], for the production of bio-crude is reported in the literature. HTL of rice straw has also been investigated in the literature [18–22]. Yuan et al. [18] studied the subcritical and supercritical liquefaction of rice straw with different solvent mixtures such as ethanol–water and 2-propanol–water. They conducted the experiments at different temperatures (260–350 °C) and solvent mixture ratios (1:9 to 5:5), and obtained 39.7 wt.% of bio-crude with 2-propanol:water at 5:5 ratio and 300 °C. Li et al. [19] investigated the HTL of rice straw by using 1,4-dioxane:water mixtures and obtained 57 wt.% bio-crude at 300 °C and equal composition of the solvent and water. Singh et al. [20] investigated the HTL of rice straw in the presence of different gases, such as N_2, O_2, and CO_2, at different temperatures, from 280 to 320 °C, at 15 min residence time, and rice-straw-to-water ratio of 1:6. The maximum bio-crude yield of 17 wt.% was obtained in N_2 ambience. The major components of the bio-crude obtained in the presence of N_2 and CO_2 ambience were phenol, guaiacol, syringol, and their derivatives. Zhou et al. [21] used Cu–Zn–Al catalyst for HTL of rice straw in presence of water–ethanol mixtures and obtained a maximum of 26.8% monomeric phenols in the bio-crude at 300 °C, 30 min, 50:50 vol.%/vol.% of ethanol:water, and 2 g of catalyst. In another study, glycerol was used as a co-solvent, along with Na_2CO_3 as the homogeneous catalyst in HTL, to produce up to 50 wt.% of bio-crude from rice straw [22]. The prospects of scale-up of HTL technology for continuous processing of feedstocks are outlined in the review by Castello et al. [23].

Supercritical alcohols in HTL of biomass possess many advantages, such as ability of donating hydrogen, better solubility, easier separation of bio-crude from kerogen-like residue, and lower corrosivity. Brand and Kim [24] studied liquefaction of cellulose, xylose, and lignin in supercritical ethanol and showed that the chemical composition of bio-crude from cellulose and xylose were quite distinct from the bio-crudes obtained via fast pyrolysis and HTL under subcritical conditions. Methanol possesses low critical temperature and pressure (239 °C, 8.09 MPa), and has established production routes from coal, natural gas, and biomass. In India, it is projected as the future fuel, with blending at up to 15% in transportation fuels. It is also expected to be available at a competitive price in India, owing to its production from coal via a two-step process involving gasification, followed by catalytic reforming of syngas [25]. Chemically, methanol is a hydrogen donor solvent, and hence, a study on its use to improve the quality of bio-crude assumes importance. This study is novel in the following respects: (1) This is the first study to report the use of methanol as a co-solvent in HTL of rice straw; (2) this study exhaustively evaluates the effect of process conditions such as temperature, reaction time, biomass-to-solvent ratio, solvent composition, and alkali catalysts on the yields and quality of bio-crude from rice straw; and (3) this study reports the energetics of the process, to evaluate the feasibility at salient operating conditions. The quality metrics include organic composition, elemental analysis and calorific value of bio-crude, and elemental composition of calorific value of bio-char.

2. Experimental

2.1. Materials and Feedstock Characterization

Rice straw was procured locally from a farm in Sullurupeta (13.7009° N, 80.0209° E), Andhra Pradesh, India. Reagents like methanol, acetone, dichloromethane (DCM), potassium hydroxide (KOH), sodium hydroxide (NaOH), and potassium carbonate (K_2CO_3) were purchased from Avra Synthesis Pvt. Ltd., India. All the reagents were used as received. Rice straw was cut into fine pieces of average size in the range of 1–1.5 cm, and stored in zip-lock covers.

Rice straw was characterized by using proximate and ultimate analyses. Proximate analysis was carried out in a thermogravimetric analyzer (TGA) (SDT Q600, T.A. Instruments, Waters GmbH, Austria), as per ASTM E1131-08 [26]. Elemental CHNS composition was obtained by using an elemental analyzer (Thermo Flash 2000, Thermo Fisher Scientific, Austria) according to ASTM D5373-08. Higher heating value (HHV) was determined by using a bomb calorimeter (IKA C2000, I.K.A., Germany) according to ASTM D5865-13. The same analytical methods were adopted to characterize bio-crude and bio-char. The total organic content in the aqueous phase was analyzed by using Karl Fischer Titrator (Metrohm-870 KF Titrino plus, Metrohm AG, Switzerland) according to ASTM D-6304. The proximate analysis, ultimate analysis, and HHV of rice straw are reported in Table 1. These characterization data are in line with the literature [27].

Table 1. Characterization of rice straw.

	Proximate Analysis (wt.%) *				Elemental Analysis (wt.%) **					HHV
	Volatile Matter	Fixed Carbon	Moisture	Ash	C	H	N	S	O	(MJ kg^{-1})
Rice Straw	66.1	15.4	5.7	12.8	37.1	5.2	0.5	0.1	44.3	12.1

* Air dried basis; ** dry basis.

2.2. HTL Reactor

HTL experiments were conducted in a custom-built stainless-steel autoclave of 1.3 L volume with 82 mm internal diameter, 25 mm thickness, and 210 mm height. The design temperature and pressure of the reactor were 400 °C and 22 MPa, respectively. The schematic of the reactor setup is provided in Figure 1. An electric coil was wrapped around its surface to provide uniform heat input. The temperature of the reaction mixture was monitored by using a K-type thermocouple, which was connected to a temperature controller. The thermocouple was placed inside the thermo-well, which was fixed to the head of the reactor vessel. A pressure gauge connected to the reactor head was used to monitor the reactor pressure, which was not controlled. Initially, the reaction mixture in the vessel was pressurized by using inert nitrogen gas. The reactor was equipped with an internal cooling coil to quickly bring down the temperature of the reactants. A four-blade turbine impeller was used to thoroughly mix the reactants. The impeller was connected through a zero-leakage magnetic drive with toque capacity of 1.96 N-m. The magnetic drive was cooled by using water circulation when the autoclave reached temperatures higher than 350 °C. The maximum rated speed of the magnetically driven stirrer was 1200 rpm. In order to withstand high pressures, graphite gaskets were used to seal the reactor and its contents. After the reaction, the reactor contents were emptied through a 10 mm diameter opening at the bottom, which was fitted with a flush valve.

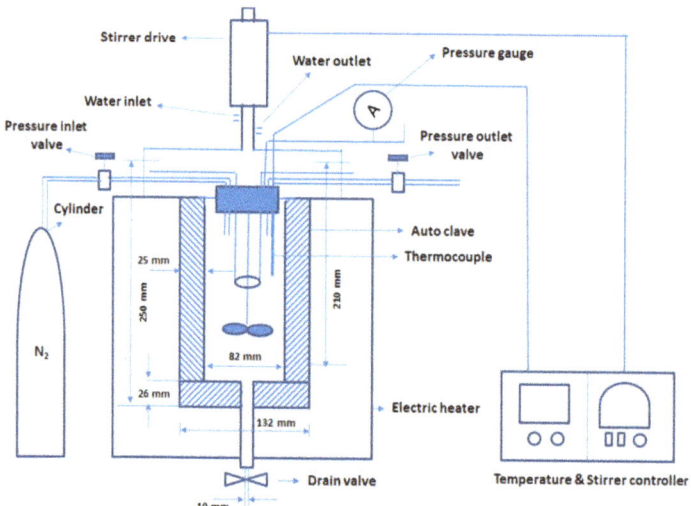

Figure 1. Schematic of the laboratory scale batch HTL reactor.

2.3. Experimental Procedure

In a typical experiment, 30 g of rice straw and 300 mL of solvent were taken in the reactor. The solvent was either water or water:methanol mixtures (75:25, 50:50, and 25:75 vol.%/vol.%). In order to investigate the effect of water content on bio-crude yield and its composition, rice straw:water composition was also varied as 1:1, 1:5, 1:10, and 1:15 (wt./wt.) in some experiments. In certain experiments, 5 wt.% of alkaline homogeneous catalysts (such as NaOH, KOH, and K_2CO_3) was mixed with water. The autoclave was sealed and then pressurized with nitrogen (N_2) gas. Typically, the initial pressure was 2–6 MPa, which was set based on the final pressure (18 MPa) and temperature (300, 325, and 350 °C) to be reached. The reactants were agitated, using the stirrer at 350 ± 25 rpm. Finally, the reactor was maintained isothermal at the set temperature for 30 and 60 min. After the reaction, the reactor was cooled down to room temperature by continuous circulation of water in the cooling coil and by means of external fan. The gases were released through the release valve, while the liquid product (or bio-crude) was separated from the solid residue (bio-char) by using a solvent-extraction process. The temperature and pressure profiles during the HTL process are depicted in Supplementary Materials Figure S1. It is evident that, in order to reach a final pressure of 18 MPa and 300 °C, the reactor was initially pressurized to ~6 MPa, using N_2. Within 50–60 min, the operating conditions were attained with a variation of ±0.5 MPa or ±5 °C around the set-point value.

The typical procedure for the separation of products is available in Chopra et al. [28]. Briefly, DCM solvent (50 mL) was used to flush the contents of the reactor, which was a viscous kerogen-like mixture containing bio-char, bio-crude, and aqueous-soluble organics. The flushed contents were collected in a beaker and were allowed to settle. The aqueous phase was then separated by decantation in a separation funnel. The DCM-soluble organic fraction was then filtered using a Whatman filter paper (pore size ~ 11 μm). The bio-char collected on the filter paper was dried at 105 °C and 24 h, and then weighed to calculate its yield. The filtrate was evaporated at 40 °C to obtain the bio-crude. DCM is a better extraction solvent over other common solvents like acetone or hexane, because of its non-polar and volatile nature, and its miscibility with many organic molecules. The use of DCM aids in maximum solubilization of the organics into bio-crude fraction, and easier separation of the aqueous phase from the organic phase. The bio-crude was then weighed to calculate its yield. The autoclave was thoroughly rinsed and washed with both DCM and acetone solvents, to eliminate traces of viscous bio-crude sticking to the reactor wall, cooling coil tubes, and stirrer blades. Most of the HTL experiments were repeated two times, while some were repeated in triplicate. The standard deviations in product yields are within 2%–5%.

The yields of bio-crude (Y_{BC}), bio-char (Y_{BCh}) and gas+aqueous fractions (Y_{G+Aq}), conversion of rice straw, and energy recovered in bio-crude (ER_{BC}) were calculated by using the following expressions:

$$Y_{BC}\ (\%) = \left(\frac{M_{BC}}{M_{RS}}\right) \times 100 \quad (1)$$

$$Y_{BCh}\ (\%) = \left(\frac{M_{BCh}}{M_{RS}}\right) \times 100 \quad (2)$$

$$Y_{G+Aq}\ (\%) = (100 - Y_{BC}\ (\%) - Y_{BCh}(\%)) \quad (3)$$

$$Conversion\ of\ rice\ straw\ (\%) = \left(\frac{M_{RS} - M_{BCh}}{M_{RS}}\right) \times 100 \quad (4)$$

$$ER_{BC}(\%) = \left(\frac{HHV_{BC} \times M_{BC}}{HHV_{RS} \times M_{RS}}\right) \times 100 \quad (5)$$

In the above expressions, M_{BC}, M_{BCh}, M_{G+Aq}, and M_{RS} denote the mass of bio-crude, mass of bio-char, mass of gas+aqueous fraction, and mass of rice straw, respectively; and HHV_{BC} and HHV_{RS} denote the HHVs of bio-crude and rice straw, respectively.

2.4. Product Characterization

The organic composition of the bio-crude samples was analyzed in a gas chromatograph/mass spectrometer (GC/MS, Shimadzu QP 2020, Shimadzu (Asia Pacific) Pte. Ltd., Singapore). The bio-crude samples were diluted in DCM solvent, and 1 µL was injected. The GC column, Rxi-5SilMS (30 m length × 0.25 mm inner diameter × 0.25 µm film thickness), was used to separate the compounds. The flow rate of helium gas (99.9995% purity) through the column was 1 mL min^{-1} with split ratio of 1:100. The injector temperature was set at 280 °C. The GC column oven temperature was initially maintained at 40 °C for 1 min, followed by a temperature ramp at 5 °C min^{-1} to 280 °C, and finally held at 280 °C for 5 min. The electron impact ionization voltage was 70 eV, and the ion-source temperature was set at 250 °C. The m/z range in the MS was 50–500 Da. Peak identification was done by comparing the mass spectra with NIST14 and Wiley08 mass spectrum libraries, and the identified compounds had a match factor greater than 85%. GC/MS peak area% (or percent selectivity) was used to quantify the various organics in the bio-crude. The organic compounds were classified according to different functional groups.

Non-condensable gases such as CO, CO_2, and H_2 from selected experiments were analyzed in a GC (Agilent 7820A, Agilent Technologies, U.S.A.) equipped with a thermal conductivity detector (TCD). The gas components were separated by using a molecular sieve-13X packed column. The carrier gas was N_2 (99.999% purity), and its flow rate through the column was 3 mL min^{-1}. The gas concentrations were measured by calibrating the column, using standard gas mixtures of different compositions.

3. Results and Discussion

3.1. Effect of Temperature and Reaction Time

Table 2 presents the mass conversion of rice straw and the yields of bio-crude, bio-char, and gas+aqueous fractions at different temperatures (300, 325, and 350 °C) and residence time periods (30 and 60 min), using water as the solvent. The mass conversion of biomass exhibits reasonable trends with the operating conditions. At any reaction time, the conversion increases with increasing temperature, as a result of the greater hydrolytic decomposition of rice straw at high temperatures. Increasing the reaction time leads to only a marginal increase in conversion of the feedstock to bio-crude and gas+aqueous fraction, which is evident by comparing the results of R1–R3 versus R4–R6. However, reaction time plays a major role in the distribution of bio-crude and gas+aqueous fractions. At 300 °C, increasing the reaction time enhances the yield of bio-crude, while at 350 °C, significant enhancement in the yield of gas+aqueous fraction is evidenced. It is also important to note that, owing to a high amount of ash in rice straw, which adds to its recalcitrance, obtaining conversions more than 80 wt.% may not be possible.

Table 2. Effect of final temperature and reaction time on conversion of rice straw and product yields, using water as solvent at rice-straw-to-water ratio of 1:10.

Expt. Code	Temperature (°C)	Reaction Time (min)	Conversion (%)	Bio-Crude Yield (wt.%)	Bio-Char Yield (wt.%)	Gas + Aqueous Yield (wt.%)
R1	300	30	75.5	10.8 ± 0.4	24.6 ± 0.5	64.6
R2	325	30	74.4	15.1 ± 0.6	25.6 ± 1.3	59.3
R3	350	30	78.6	12.9 ± 0.7	21.4 ± 0.8	65.7
R4	300	60	75.2	12.3 ± 1.5	24.8 ± 1.8	62.9
R5	325	60	77.1	9.2 ± 0.5	22.9 ± 0.7	67.9
R6	350	60	80.0	3.1 ± 1.2	20.0 ± 0.3	76.9

The effect of temperature on bio-crude yield is evident from Table 2 at similar processing times. For example, after 30 min of processing time, the bio-crude yields followed the trend 15.1 wt.% (325 °C) > 12.9 wt.% (350 °C) > 10.8 wt.% (300 °C). Increasing the temperature from 300 to 350 °C at 60 min processing time led to a significant decrease in the yield of bio-crude (12.3 to 3.1 wt.%), with a concomitant increase in gas+aqueous yield (62.9 to 76.9 wt.%) and only a minor drop in bio-char yield. This suggests that thermal decomposition and degradation of the organic components are predominant at high temperatures. However, at a lower temperature of 300 °C, increasing the reaction time to 60 from 30 min leads to an enhancement in bio-crude yield, without significant change in conversion. This shows that temperature and processing time do have a combined effect on bio-crude yields. At 30 min processing time, a maximum bio-crude yield of 15.1 wt.% was achieved at 325 °C, while at 60 min, a maximum bio-crude yield was obtained at 300 °C. Run-dong et al. [29] studied HTL of rice stalks at various reaction time periods (30–120 min) and temperatures (250–375 °C), in the presence of ethanol. At 250 °C, a maximum bio-crude yield of 48 wt.% was obtained at an optimal reaction time of 90 min, whereas the highest bio-crude yield, of 55 wt.%, was obtained at 60 min and 325 °C. This shows that, at a high temperature, a shorter residence time may be sufficient to achieve high bio-crude yield. However, the optimum operating condition is not only determined by bio-crude yield, but also by other bio-crude quality markers, such as HHV, and elemental composition (H/C and O/C).

Table 3 depicts the elemental composition and HHV of bio-crude and bio-chars obtained from experiments R1–R6. The HHVs of bio-crude were calculated by using the correlation developed by Francis and Lloyd [30]: HHV (MJ kg^{-1}) = (34 [%C] + 124.3 [%H] + 6.3 [%N] + 19.3 [%S] − 9.8 [%O])/100. As explained earlier, 350 °C–60 min corresponds to a severe condition that leads to low carbon content in bio-crude, in addition to low yield. As a result, the HHV of bio-crude is also the lowest as compared to other experiments performed in this study. It is important to note that the quality of bio-crude is better from experiment R4, corresponding to 300 °C–60 min. This is evident from the high carbon (77.6 wt.%) and low oxygen content (15.3 wt.%). The HHV of the bio-crude obtained at this condition is also high (35 MJ kg^{-1}), and it is in accordance with the oxygen content. In order to further ascertain which operating condition is the optimum, the energy recovery in bio-crude was assessed. Energy recovery in bio-crude depends on both bio-crude yield and its HHV, according to Equation (5). At 300 °C–60 min, the energy recovery of ~36% is due to both high HHV of bio-crude (35.3 MJ kg^{-1}) and its reasonable yield (12.3 wt.%). Moreover, the hydrogen content in bio-crude is maximum (8.4 wt.%) at this condition, and the atomic H/C and O/C ratios are 1.31 and 0.14, respectively. The HHVs of bio-crudes observed in this study are in line with that reported in the HTU® (Hydrothermal Upgradation) process of Shell Research Laboratory for a variety of biomass feedstocks [7]. A preliminary analysis of the non-condensable gases produced at this condition was performed, using GC–TCD. The major gases were CO_2 (74 vol.%), CO (18 vol.%), CH_4 (6 vol.%), and H_2 (2 vol.%) on N_2-free basis. The high amount of CO_2 generated can be attributed to the decarboxylation reactions, which are dominant than the CO removal through decarbonylation during HTL. When compared to fast pyrolysis bio-oil from rice straw, which contains 42%–49% of oxygen [31,32], HTL bio-crude from rice straw at the optimal processing condition contains three times lower oxygen (14.2%), which demonstrates its superior quality. Therefore, further experiments involving different rice straw:water ratios, and water:methanol ratios were performed at a final temperature of 300 °C and 60 min residence time.

Table 3. Effect of final temperature and reaction time on elemental composition of bio-crude and bio-char, HHV of bio-crude and bio-char, and energy recovered in bio-crude. The rice-straw-to-water ratio is 1:10 wt./wt.

Expt. Code	Temp. (°C)	Time (min)	Elemental Composition of Bio-Crude (wt.% db)				HHV of Bio-Crude (MJ kg^{-1})	Elemental Composition of Bio-Char (wt.% db)				HHV of Bio-Char (MJ kg^{-1})	ER_{BC} (%)
			C	H	N	O*		C	H	N	O#		
R1	300	30	72.9	7.9	0.3	18.9	32.7	51.6	4.5	0.4	30.7	20.1	29.4
R2	325	30	73.5	7.0	0.5	19.0	31.9	46.6	3.4	0.3	36.9	16.5	39.8
R3	350	30	69.2	6.9	0.6	23.2	29.8	53.4	3.8	0.4	29.6	20.0	31.7
R4	300	60	77.1	8.4	0.4	14.1	35.3	51.3	4.3	0.2	31.4	19.8	35.9
R5	325	60	73.3	7.0	0.4	19.3	31.7	39.5	3.2	0.3	44.2	13.1	24.1
R6	350	60	65.7	6.4	0.5	27.4	27.6	49.7	3.4	0.4	33.7	17.8	7.1

* %O = 100-(%C + %H + %N), # %O = 100-(%C + %H + %N + %Ash). db: dry basis. ER: energy recovered in bio-crude (%).

The elemental composition of bio-char from various experiments is depicted in Table 3. It is clear that most of the oxygen present in the feedstock is transferred to the bio-char in HTL experiments. This is reflected in the high oxygen content (28–44 wt.%) in bio-char. The carbon content in bio-chars is lower than that present in bio-crude, but certainly better than rice straw feedstock. As a result, the HHVs of majority of bio-chars are higher (17–21 MJ kg^{-1}) than that of the rice straw (12.1 MJ kg^{-1}). The production of bio-char can be attributed to both fixed carbon present in the feedstock and the repolymerization reactions occurring at long residence periods during the process. The repolymerization reactions are particularly important at high pressures and low temperatures. At higher temperatures, there is a possibility of the oxygenates and hydrocarbons to get converted to non-condensable gases.

3.2. Effect of Rice-Straw-to-Water Ratio

Figure 2 depicts the product yields when the mass ratio of rice straw to water is varied at the optimal condition of 300 °C–60 min. It is evident that the bio-crude yield increases from 5.7 to 12.3 wt.% when the rice-straw-to-water ratio is increased from 1:1 to 1:10 wt./wt. The further increase in water content decreases the bio-crude yield to 5.8 wt.%. Generally, a high amount of the solvent leads to better extraction of the organics produced during the HTL process by the denser solvent medium [33]. This is supported by the trend followed by bio-char yield: 35.5 wt.% (1:1 wt./wt.) > 25 wt.% (1:5 wt./wt.) ≈ 24.8 wt.% (1:10 wt./wt.) > 23.9 wt.% (1:15 wt./wt.). However, rice-straw-to-water ratio of 1:10 wt./wt. is found to be the optimal condition for maximum bio-crude production. The HHV of bio-crude increased from 33.9 to 35.3 MJ kg^{-1} with the increase in rice-straw-to-water ratio from 1:1 to 1:10 wt./wt., and then decreased to 33.7 MJ kg^{-1} with further increase, up to 1:15 wt./wt. This is attributed to the increase of oxygen content in bio-crude to 15.3 from 14.1 wt.% as the water content was increased. Maximum gas+aqueous yield of 70 wt.% was obtained with 1:15 wt./wt. of rice straw: water, which shows that small-molecule oxygenates of the bio-crude were probably converted to gases. Moreover, it is possible that high water content in the feedstock can solvate the small molecule oxygenates, and so they get transferred to the aqueous phase. Therefore, 1:10 wt./wt. is verified to be the optimal rice straw:water composition.

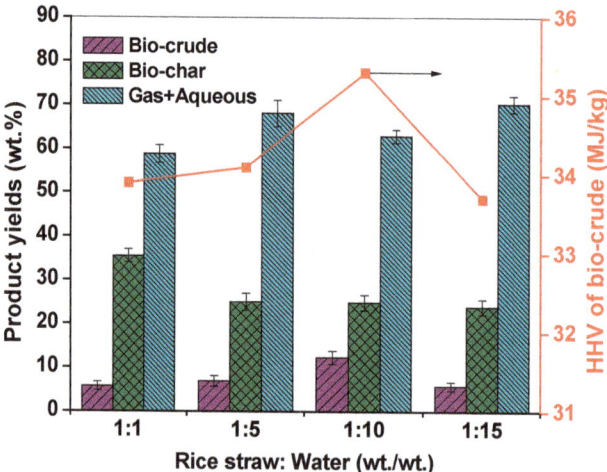

Figure 2. Effect of rice-straw-to-water ratio on product yields and HHV of bio-crude at a final temperature of 300 °C and 60 min.

3.3. Influence of Methanol as a Co-Solvent

Co-solvents play a significant role in enhancing the production of bio-crude because of the strong and specific interactions with biomass intermediates. Co-solvents are found to be more effective than an individual solvent for bio-crude production from biomass via HTL [9]. In fact, the effect is due to both physical and chemical factors. The critical point of methanol (239 °C, 8.09 MPa) is low as compared to that of water (374 °C, 22.1 MPa). The critical temperature and pressure values of various methanol:water mixtures (in vol.%/vol.%) were calculated by using ASPEN plus V8.6 software, and they followed the trend 25:75 (352.7 °C, 19.9 MPa) > 50:50 (325.5 °C, 17 MPa) > 75:25 (289.5 °C, 13.3 MPa). The dielectric constant of methanol is low as compared to that of water. Therefore, methanol can readily dissolve the high molecular weight organics generated from rice straw decomposition. Moreover, at 50:50 and 75:25 compositions of methanol:water, supercritical condition prevails at the operating condition. Therefore, these are expected to significantly affect the product yields and their composition. The chemical effect of methanol is its ability to donate hydrogen. Pedersen et al. [34] reported that radicals formed from hydrogen-donating solvents can act as radical quenching mediators and prevent the lignin-derived radicals from participating in repolymerization reactions. Other studies have reported that alcohols help to dissolve and stabilize the intermediates. This reduces their rate of repolymerization, which eventually enhances the bio-crude yield [35,36]. The hydrogen donor solvents enhance the hydrogenolysis reactions of biomass. In their absence, the formed radicals from rice straw can recombine to form high-molecular-weight compounds, which eventually increase the yield of bio-char.

Figure 3 depicts the yields of product fractions and the HHVs of bio-crudes obtained at different water:methanol compositions. Clearly, the presence of methanol as co-solvent improves the yield of bio-crude, which follows the trend 31.7 wt.% (50:50 water:methanol) > 16.4 wt.% (25:75) > 14.4 wt.% (75:25) > 12.3 wt.% (100:0). The bio-char yield follows a clear trend with water:methanol composition, as follows: 20.76 wt.% (25:75) > 18.5 wt.% (50:50) > 14.9 wt.% (75:25).

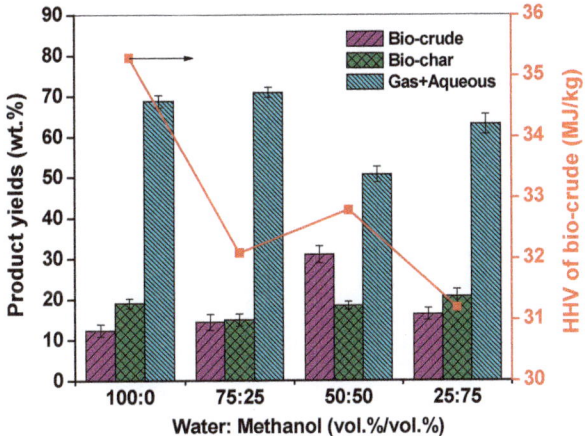

Figure 3. Effect of water-to-methanol ratio on product yields and HHV of bio-crude at a final temperature of 300 °C, 60 min reaction time, and rice-straw-to-solvent ratio of 1:10 wt./wt.

In order to ascertain the quality of bio-crudes obtained by using methanol as co-solvent, the elemental composition and HHVs were analyzed. From Table 4, it is clear that the oxygen content in bio-crude is, in fact, higher (17.8–22.3 wt.%) as compared to that obtained with only water as the solvent. The carbon content in bio-crude is lower (69–72 wt.%) as compared to that when methanol was not used. With equal composition mixture of water:methanol (experiment R11), the oxygen content is reasonably low (17.8 wt.%), which leads to better HHV of bio-crude at this condition. Contrary to the expectation, there is no significant improvement in hydrogen content of the bio-crude, while oxygen is improved. This indicates the participation of methanol in reactions with the hydrolysis intermediates from biomass during HTL. This is evident in the formation of high amounts of oxygenated compounds, like esters in bio-crude, which is discussed in Section 3.5. The HHVs of bio-crudes are in line with the oxygen content in them. The HHV recorded with 50:50 (vol.%/vol.%) water:methanol was 32.8 MJ kg^{-1}. Li et al. [37] conducted HTL of sludge in supercritical ethanol and showed that oxygen content in bio-crude passed through a maxima as ethanol:water content was varied. They attributed this to the competition between hydrogen donor reduction of oxygen, and CO–CO_2 reduction of oxygen. In this study, oxygen content is minimum with equal composition mixture of methanol:water, which then increases with the addition of methanol. This shows that more methanol in the mixture possibly leads to hydrogen donor reduction of oxygen, which leads to dehydration. Thus, a lower amount of oxygen from biomass is lost as water, as compared to a higher amount of oxygen loss in the form of CO_2 and CO.

The bio-char obtained from experiments conducted with methanol (R10–R12) contained a high amount of oxygen and low amount of carbon in them. This is reflected in their poor HHVs (<15.7 MJ kg^{-1}). This shows that, besides getting converted to oxygenates in bio-crude, a significant fraction of methanol is involved in the formation of bio-char. This occurs by the formation of high-molecular-weight oxygenated polymerized fragments. The analysis of non-condensable gases collected at the end of the HTL process conducted with equal composition of water:methanol (experiment R11) showed 40 vol.% of CO_2, 46.7 vol.% of CO, and 13.3 vol.% of H_2. As compared to experiment R4 conducted with only water as solvent, the CO_2 production is significantly low in this experiment. Methanol decomposition is reported to occur via two pathways. In one pathway, methanol reacts with water under HTL conditions, to form CO_2 and H_2. In another pathway, methanol initially decomposes, to form formaldehyde and H_2. Formaldehyde further decomposes to CO and H_2 [38]. Finally, CO can get converted to CO_2 by reacting with water. The observed gas composition demonstrates that the second pathway is favored when methanol is added as a co-solvent.

Table 4. Effect of water-to-methanol ratio and catalysts on elemental composition of bio-crude and bio-char, and HHVs of bio-crude and bio-char at final process temperature of 300 °C, reaction time of 60 min, and rice-straw-to-solvent ratio of 1:10 wt./wt.

Expt. Code	Water: Methanol (Vol.%/Vol.%)	Elemental Composition of Bio-Crude (wt.% db)				HHV of Bio-Crude (MJ kg^{-1})	Elemental Composition of Bio-Char (wt.% db)				HHV of Bio-Char (MJ kg^{-1})
		C	H	N	O *		C	H	N	O #	
R10	75:25	71.5	7.7	1.6	19.2	32.1	44.4	3.4	1.7	37.7	15.7
R11	50:50	72.7	7.8	1.7	17.8	32.8	36.4	3.1	1.3	46.4	11.7
R12	25:75	69.1	7.7	2.3	20.9	31.2	30.8	2.7	1.1	52.6	8.7
R13	50:50 [1]	71.7	7.7	2.3	18.3	32.3	27.4	2.7	1.0	56.1	7.2
R14	50:50 [2]	73.5	7.9	2.4	16.2	33.3	23.4	2.3	0.8	60.7	4.8
R15	50:50 [3]	73.8	7.7	2.3	16.2	33.2	23.3	2.3	0.8	60.8	4.9

* %O = 100-(%C + %H + %N), # %O = 100-(%C + %H + %N + %Ash), [1] KOH (5 wt.%), [2] NaOH (5 wt.%), and [3] K$_2$CO$_3$ (5 wt.%) were employed as catalysts. In R10, R11, and R12, no catalyst was used.

3.4. Role of Alkali Catalysts

The addition of homogeneous catalysts, such as alkali salts, is known to improve the bio-crude yield in the HTL process, owing to the inhibition of dehydration reaction of the carbohydrate fraction of the biomass [7]. This is particularly evident at high pH values of the mixtures upon the addition of alkali. Moreover, hydrothermal gasification and water–gas shift reactions are promoted by the alkali catalysts [7]. Alkali catalysts such as KOH, K$_2$CO$_3$, Na$_2$CO$_3$, and NaOH have been tested for HTL of agro residues, forest wastes, and municipal solid waste mixtures [39,40]. In this study, 5 wt.% of KOH, K$_2$CO$_3$, and NaOH was used as catalysts for HTL of rice straw at 300 °C–60 min, rice-straw-to-solvent ratio of 1:10 (wt./wt.), and water-to-methanol ratio of 50:50 (vol.%/vol.%). Figure 4 depicts the product yields and the HHVs of bio-crudes. It is evident that the use of alkali catalysts significantly improved the bio-crude yield, which follows the trend KOH (39.9 wt.%) > NaOH (34.5 wt.%) > K$_2$CO$_3$ (33.8 wt.%) > no catalyst (31.7 wt.%). Among the three catalysts, K$_2$CO$_3$ was most effective for overall mass conversion of rice straw to bio-crude and gas+aqueous fractions. The maximum conversion of 86.9% was obtained with K$_2$CO$_3$, followed by 84.6% for NaOH, 83.5% for KOH, and 81.6% without any catalyst.

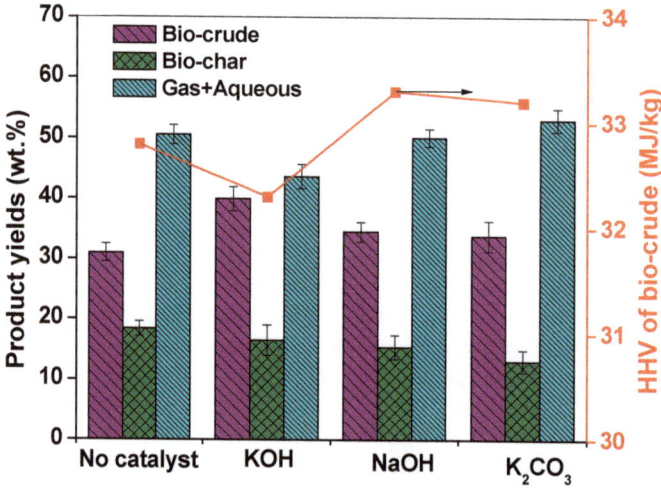

Figure 4. Effect of alkali catalysts on product yields and HHV of bio-crude at 300 °C, 60 min, rice straw:solvent ratio of 1:10 wt./wt., catalyst quantity of 5 wt.%, and 50:50 (vol.%/vol.%) methanol:water.

Generally, carbonates are shown to be effective as catalysts for the HTL process, compared to hydroxides. The mechanism of action of K_2CO_3 is proposed to follow the reaction sequence given by [7]:

$$K_2CO_3 + H_2O \rightarrow KHCO_3 + KOH$$

$$KOH + CO \rightarrow HCOO^- K^+ \text{ (formate salt)}$$

$$HCOOK + H_2O \rightarrow KHCO_3 + H_2$$

$$2KHCO_3 \rightarrow H_2O + K_2CO_3 + CO_2$$

The high yield of gas+aqueous (53.1 wt.%) produced using K_2CO_3 substantiates its effectiveness in improving the conversion of rice straw via hydrolytic decomposition of carbohydrates and lignin components. The reported gas+aqueous yields are in line with the value reported by Zhu et al. [41]. Even though the bio-crude yield was high with KOH, the oxygen content is lower (16 wt.%) and carbon content is higher (73–74 wt.%) in bio-crudes produced using NaOH and K_2CO_3. As a result, the HHVs of bio-crudes are slightly better (33 MJ kg^{-1}) with NaOH and K_2CO_3 catalysts as compared to that from KOH (32 MJ g^{-1}). The molar H/C and O/C content in bio-crudes vary in the range of 1.25–1.30 and 0.17–0.19, respectively. From the elemental composition of bio-char, it is clear that the use of these alkali salts, along with methanol co-solvent, significantly improves the oxygen retained in the bio-char. The high oxygen (56–60 wt.%) and low carbon (23–27 wt.%) lead to low HHVs of bio-chars. Table 5 presents a comparison of bio-crude yields and their salient properties from rice straw, using different solvent systems. It is evident that the yield of bio-crude obtained using methanol as co-solvent is comparable with that using 2-propanol. Importantly, the HHV of bio-crude is better and oxygen content in it is lower with the methanol–water system as compared to 2-propanol–water and ethanol–water systems. This substantiates the positive effect of methanol as a co-solvent for HTL of rice straw.

Table 5. Comparison of bio-crude yield and its properties reported in the literature with this study.

Temp. (°C)	Rice Straw (g)	Solvent (mL)	Solvent: Water (Vol./Vol.)	Solvent	Bio-Crude Yield (wt.%)	HHV (MJ kg^{-1})	O (wt.%)	C (wt.%)	Ref.
300	20	300	-	Water	29.1	27.2	45.3	47.4	Yuan et al. [18]
	20	300	-	Ethanol	13.0	33.5	-	-	
	20	300	1:1	Ethanol–water	38.4	29.7	-	-	
	20	300	-	2-Propanol	15.1	35.8	18.5	71.3	
	20	300	1:1	2-Propanol–water	39.7	30.7	27.3	63.8	
300	30	300	-	Water	12.3	35.3	14.1	77.1	This study
	30	300	1:1	Methanol–water	36.8	32.8	17.8	72.7	
	30	300	1:1	Methanol–water+KOH	40.0	32.3	18.3	71.7	

3.5. Bio-Crude Composition

Bio-crudes from selected experiments were analyzed, using GC/MS, to analyze the variation of organic composition with respect to the process parameters. The various organic compounds present in the bio-crude were classified according to the functional groups, namely simple oxygenates, cyclo-oxygenates, aliphatic hydrocarbons, phenolics, other aromatics, indene derivatives, furan derivatives, and esters. The category of other compounds includes N-containing organics and those that were obtained with low match factor in GC/MS. Simple oxygenates were mostly linear-chain-substituted ketones, while cyclo-oxygenates contained mostly cylopentenone and alkyl-substituted cyclopentenones. Aromatics included primarily non-phenolic compounds like aromatic hydrocarbons and aromatic oxygenates. Figure 5 depicts the selectivities to functional groups when 1:10 wt./wt. of rice straw:water was used at various conditions. The major product groups were found to be cyclo-oxygenates, phenolics, and other aromatics. It is evident that, with the increase in residence time of the process, from 30 to 60 min, the selectivity to phenolic compounds decreased slightly. The marginal decrease in selectivity to phenolics from 53.7% (30 min) to 50.4% (60 min) at

300 °C, and 49.9% (30 min) to 49.3% (60 min) at 325 °C can be attributed to simultaneous formation of phenolic compounds from the lignin fraction at longer processing time and the decomposition of phenols to aromatic hydrocarbons via dehydration and demethoxylation pathways. Moreover, a major reduction in selectivity to cyclo-oxygenates accompanied by an increase in the formation of non-phenolic aromatic compounds were observed. In fact, the combined selectivity to non-phenolic aromatics and indene derivatives increased significantly as the processing time was increased. For example, at 300 °C, it increased from 6.6% to 21.2%, while at 325 °C, it increased from 6.6% to 18.8%.

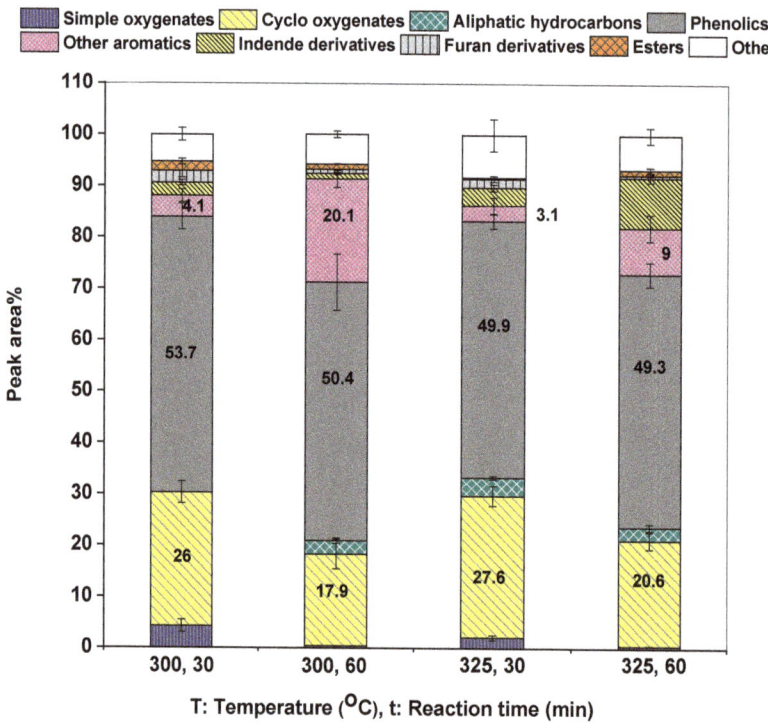

Figure 5. Selectivity to various functional groups in bio-crude obtained at rice-straw-to-water ratio of 1:10 wt./wt., at different operating conditions.

The major phenolic compounds in the bio-crude were phenol, 2-methylphenol (o-cresol), 4-ethylphenol, catechol, guaiacol, 4-ethylguaiacol, alkyl-substituted benzenediols, hydroxyl acetophenone, and syringol. Phenolic compounds are derived from the decomposition of lignin, mainly via hydrolytic cleavage of α-O-4 and β-O-4 linkages [42]. Pelzer et al. [43] proposed a mechanism of acidolysis of lignin model compounds, which involved nucleophilic attack of water on α-carbon that led to α-O-4 aryl-ether bond cleavage. This mechanism was commensurate with the protonation of ether oxygen. Similar reactions in the HTL process are responsible for the formation of observed products.

An increase in processing time clearly decreases the selectivity to cyclo-oxygenates, which can be attributed to their decomposition to form non-condensable gases, like CO and CO_2, in addition to other low molecular oxygenates, which were not detected in the gas phase in this study. For example, at 300 °C, the decrease was from 26% (30 min) to 17.9% (60 min), while it was 27.6% (30 min) to 20.6% (60 min) at 325 °C. The major cyclo-oxygenates were cyclopentanone, 2-cyclopentene-1-one, 3-methyl-2-cyclopentene-1-one, 2,3-dimethyl-2-cyclopentene-1-one, 2,3,4-trimethyl-2-cyclopentene-1-one, and 2-acetonylcyclopentanone. The cyclopentanones are mainly produced by the decomposition of the pentanose sugars present in the hemicellulose fraction of biomass. A few cyclohexanone derivatives were also observed in the bio-crude, and their formation can be traced to the successive hydrolytic dehydration of hexose sugars from the cellulose fraction. Interestingly, the decrease in production of cyclo-oxygenates at 60 min is accompanied by an increase in production of aromatic compounds such as methoxy-substituted benzenes, naphthalenol, naphthol, hydroxy-substituted naphthalene derivatives, and methyl- and hydroxy-substituted indanone. This indicates that, at long residence times, bimolecular condensation reactions of cyclo-oxygenates with low-molecular-weight compounds are prevalent during HTL.

The temperature effect is not significant for the production of these compounds in the bio-crude at similar processing periods. For example, at 60 min, the selectivities to phenolic compounds are comparable (50.4% at 300 °C vs. 49.3% at 325 °C), while the selectivities of total aromatics (combined non-phenolic aromatics and indene derivatives) are only slightly different (21.2% at 300 °C vs. 18.8% at 325 °C). Similarly, at 30 min, the selectivities to cyclo-oxygenates (26% vs. 27.6%) and total aromatics (6.6% at 300 and 325 °C) are very much comparable. Therefore, the effect of increase in temperature, at least in the range investigated in this study, is minimal on composition of major functional groups in bio-crude. Figure 6 depicts the structures of the salient products produced from HTL of rice straw in presence of water as the solvent. Supplementary Tables S1–S4 provide the list of classified organic compounds detected in the bio-crude at different conditions, along with their composition. A number of compounds identified in this study are in agreement with Liu et al. [44], who carried out HTL of walnut shells in the presence and absence of alkali.

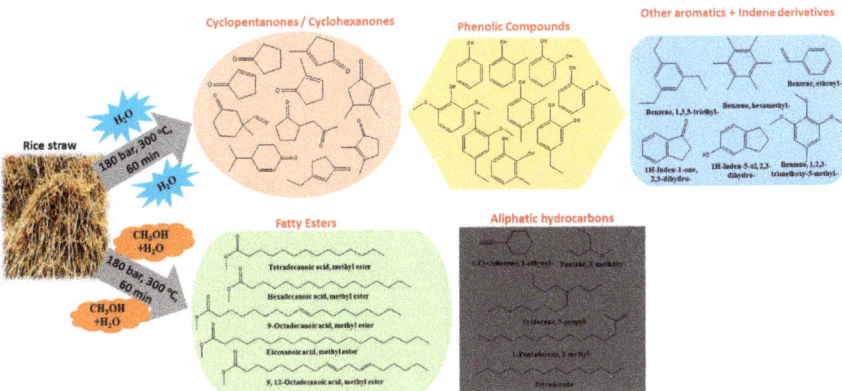

Figure 6. Scheme depicting the typical products produced from HTL of rice straw at different conditions.

The effect of water:methanol and alkali catalysts on bio-crude composition is shown in Figure 7a. The detailed lists of classified organics under each product group at different water:methanol volume ratios are available in Supplementary Tables S5–S7. It is evident that the addition of methanol to water significantly alters the organic distribution in the bio-crude. Esters, mostly methyl esters, emerge as the major organic compounds. Compared to the case with only water as the solvent, the addition of methanol enhances the production of methyl esters, possibly via esterification/methanolysis reactions of the carboxylic functional groups present in the biomass intermediates, which is similar to reported results [10,45]. The subcritical/supercritical medium of methanol–water can catalyze these reactions, owing to the strong acidic–basic nature. The selectivity to esters in the bio-crude follows the trend 72.9% (25:75 vol.%/vol.% water:methanol) > 59.6% (75:25) > 51.5% (50:50). Interestingly, a similar trend is followed by the oxygen content in the respective bio-crudes (Table 4). This validates the high HHV of bio-crude when an equal composition mixture of water:methanol is used as the solvent system (Figure 3). The major esters include octadecanoic acid–methyl ester, hexadecanoic acid–methyl ester, and tetradecanoic acid–methyl ester. Generally, the high selectivity to esters is accompanied by lower selectivities to cyclo-oxygenates and phenolic compounds. More importantly, it is worthwhile to mention that with 25:75 water:methanol mixture, the operating condition of 300 °C–18 MPa is supercritical, while for 75:25 and 50:50 water:methanol mixtures, the operating condition is still subcritical and near critical, respectively. This can lead to significant differences in solubility of the components in bio-crude, which has resulted in the observed distribution. The production of simple/linear oxygenates, aliphatic hydrocarbons, furan derivatives, and indene derivatives was negligible in the bio-crude when methanol was added as a co-solvent. Li et al. [37] observed ethyl esters and butyl esters of long-chain carboxylic acids (C15–C20) and fatty acids (mainly octadecanoic acid) as the major components in the bio-crude from sludge in the presence of supercritical ethanol. This supports the organic distribution in bio-crude observed in this study. Figure 6 depicts the structures of the salient esters and aromatic compounds produced from rice straw when methanol was used as the co-solvent.

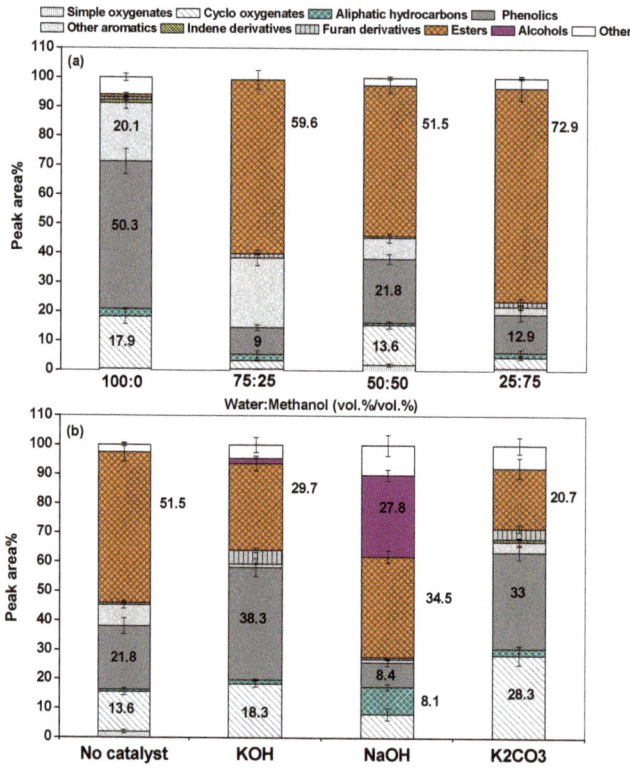

Figure 7. Selectivity to various functional groups in bio-crude obtained at (**a**) different water-to-methanol ratios and (**b**) different alkali catalysts at water:methanol of 50:50 (vol.%/vol.%). The final temperature is 300 °C, reaction time is 60 min, and rice-straw-to-solvent ratio is 1:10 wt./wt.

Figure 7b depicts the organic distribution in bio-crude from rice straw when the alkali catalysts were used in the presence of an equal composition mixture of methanol:water. The detailed lists of products in the bio-crude classified according to functional groups are available in Supplementary Tables S8–S10. It is evident that the selectivity to esters is greatly reduced upon the addition of the alkali catalysts, which shows that these promote hydrolysis reactions over esterification, which is due to the effect of methanol in the reaction mixture. The selectivities to esters follow the trend: no catalyst (51.5%) > NaOH (34.5%) > KOH (29.7%) > K_2CO_3 (20.7%). Significant production of phenolic compounds is observed in the presence of KOH (38.3%), followed by K_2CO_3 (33%) and un-catalyzed system

(21.8%). The addition of NaOH, interestingly, favors the production of alcohols, mainly dodecanol and tetradecanol, which shows that hydroxylation is favored in presence of NaOH. The production of long-chain hydrocarbons is also favored (~8% selectivity in presence of NaOH), but the phenolics were greatly suppressed. For HTL of woody biomass, it is reported that potassium salts exhibit better catalytic activity as compared to sodium salts, which follow the trend K_2CO_3 > KOH > Na_2CO_3 > NaOH [7]. Importantly, the role of alkali salt depends on the biomass type to a certain extent, and the solvent medium to a great extent. Therefore, the observed variations in product distribution are specific to the reaction mixture considered in this study. Between KOH and K_2CO_3, the former is found to be effective for the hydrolytic cleavage of lignin to produce phenolics, while the latter exhibits relatively better activity for hemicellulose hydrolysis to produce more of cyclo-oxygenates.

3.6. Process Energy Balance

In order to evaluate the energy consumed by various streams and the process as a whole, a detailed energy balance of the process for experiments R4, R11, and R13 was performed. These experiments were performed at a final temperature of 300 °C and residence time of 60 min. The percent energy recovery and energy-consumption ratio were determined according to the following expressions [46,47]:

$$\text{Energy recovery (ER) (\%)} = \frac{E_P}{E_F + Q_S} \times 100 \quad (6)$$

$$\text{Energy consumption ratio (ECR)} = \frac{E_{HTL}}{E_{BC}} \quad (7)$$

where E_P, E_F, E_{HTL}, E_{BC}, and Q_S denote the total energy of products, energy of the feedstock, heat required for the HTL process, energy recovered in bio-crude, and heat of solvent, respectively. These terms are defined as follows [46,47]:

$$E_P = (M_{BC} \times HHV_{BC}) + (M_{BCh} \times HHV_{BCh}) + (M_G \times HHV_G) \quad (8)$$

$$E_F = (M_{RS} \times HHV_{RS}) \quad (9)$$

$$E_{HTL} = \frac{(Q_S + Q_{RS}) \times (1 - \eta_h)}{\eta_c} \quad (10)$$

$$E_{HTL} = \frac{((m_S \times C_{p,S} \times \Delta T) + (m_{RS} \times C_{p,RS} \times \Delta T)) \times (1 - \eta_h)}{\eta_c} \quad (11)$$

$$E_{BC} = (M_{BC} \times HHV_{BC}) \quad (12)$$

In the above expressions, m_S, m_{RS}, $C_{p,S}$, $C_{p,RS}$, and ΔT denote the mass of solvent, mass of rice straw, specific heat capacity of solvent, specific heat capacity of rice straw, and temperature difference, respectively. Q_{RS} and Q_S denote the heat required to raise the temperature of rice straw and solvent, respectively. When water–methanol mixtures were used as solvents, Q_S was calculated by mixture rule. The factors η_h and η_c in Equation (11) denote the efficiencies of heat recovery and combustion energy, respectively, and correspond to the scenario of bio-crude combustion. The values of these are modestly assumed as 0.5 and 0.7, respectively [47]. The sample calculations are provided in Supplementary S11.

Table 6 depicts the energy values for three experiments. It is evident that the energy recovery increases on addition of methanol as co-solvent, which is primarily attributed to the increase in yield of bio-crude. Nearly two times improvement in energy recovery is recorded with experiment R13 as compared to R4. This shows that the addition of methanol and the homogeneous catalyst play a vital role. The ECR provides an estimate of the effectiveness of the HTL process compared to the energy that can be recovered by combustion of bio-crude. ECR value greater than unity signifies that the process consumes more energy than that generated in the bio-crude, and vice versa when ECR is lower than unity. Therefore, ECR < 1.0 is preferred. It is evident from Table 6 that, with only water as the solvent,

ECR is 2.03. This shows that the process may not be favorable from an energy perspective. However, with the addition of methanol and the catalyst, a net-positive energy balance is obtained. It can be concluded that both the processes (addition of methanol and catalyst) are energetically feasible. A particular advantage of adding methanol as a co-solvent is reflected in the low energy required to raise the temperature of the slurry to the required temperature, as compared to using only water as the solvent. In addition to the energy advantage, the addition of co-solvent also improves the yield of the bio-crude, which is reflected in energy recovery, which is nearly two times higher as compared to that using only water. These are comparable with HTL of microalgae and defatted microalgae, and certainly better than that of pyrolysis bio-oil [47]. An ensemble of the results demonstrates that HTL, using methanol as a co-solvent, is a promising thermochemical technology to convert the abundantly available waste rice straw to valuable bio-crude with low oxygen content and high HHV.

Table 6. Energy calculations to determine energy recovery and ECR in HTL of rice straw at 300 °C, 60 min, and rice straw:solvent ratio of 1:10 wt./wt.

	Water	Water:Methanol 50:50 (Vol.%/Vol.%)	Water:Methanol 50:50 (Vol.%/Vol.%) + KOH
Energy content of rice straw (E_F) (kJ)	363	363	363
Heat of rice straw (Q_{RS}) (kJ)	6.93	6.93	6.93
Heat of water (Q_{H2O}) (kJ)	362.62	181.30	181.30
Heat of methanol ($Q_{Methanol}$) (kJ)	-	32.65	32.65
Heat required for HTL process (E_{HTL}) (kJ)	263.96	157.80	157.80
Total energy of products (E_P) (kJ):	322.62	424.50	478.34
Energy recovered in bio-crude (kJ)	130.26	305.00	387.83
Energy recovered in bio-char (kJ)	137.64	64.75	35.79
Energy recovered in gases (kJ)	54.72	54.72	54.72
$E_F + Q_{solvent}$ (kJ)	725.62	576.96	576.96
Energy recovery (ER) (%)	44.5	73.6	82.9
Energy consumption ratio (ECR)	2.03	0.52	0.41

4. Conclusions

Through a series of systematic experiments, a better-quality bio-crude was obtained from rice straw at 300 °C, 18 MPa, and a 60 min processing time, using only water as the solvent at rice straw:water ratio of 1:10 (wt./wt.). At this condition, the bio-crude yield was 12.3 wt.%, HHV was 35.3 MJ kg^{-1}, oxygen content was 14.2 wt.%, and energy recovery was 36%. The major organic constituents in the bio-crude were phenolic compounds, cyclo-oxygenates like cyclopentenone and its derivatives, and other aromatics. The selectivities to phenolic compounds, cyclopentenone, and its derivatives and aromatics at this condition were 50%, 18%, and 21%, respectively. In order to further improve the yield and quality of bio-crude, methanol was used as a co-solvent. The yields of bio-crude increased up to 36.8 wt.% with the addition of methanol in equal volume ratio with water. The oxygen content and HHV of bio-crude were 17.8 wt.% and 32.8 MJ kg^{-1}, respectively. Esters emerged as the major constituents of bio-crude, owing to the enhancement of esterification reactions under acid/base-catalyzed conditions in presence of methanol:water. The addition of alkali catalysts further improved the yield of bio-crude, with a maximum yield of ~40 wt.% obtained using 5 wt.% KOH. The catalysts promoted both hydrolytic cleavage of biomass components to produce phenolics and cyclo-oxygenates, and esterification reactions to produce methyl esters. Energy recovery was calculated to be in the range of 70%–80% for HTL conducted in presence of methanol:water. The energy-consumption ratio was lesser than unity for the methanol:water system, suggesting that the process produced more energy in the bio-crude than it consumed.

This study proves that recovery of valuable esters may be targeted from HTL bio-crude for their use in various industries, such as soap manufacturing, cosmetics, food, detergents, and chemicals. Moreover, an analysis of the energy-consumption ratio for the presented scenarios in a continuous

HTL process with energy integration among multiple unit operations, and a detailed study on the techno-economic aspects are essential to establish the commercial viability of the HTL technology for biomass agro residue valorization.

Supplementary Materials: The following are available online at http://www.mdpi.com/1996-1073/13/10/2618/s1. Figure S1: Temperature and pressure profiles from selected experiments performed at rice-straw-to-water ratio of 1:10 wt./wt. Tables S1–S10: Detailed composition of bio-crude from experiments R1, R2, R4, R5, R10, R11, R12, R13, R14 and R15, respectively. S11: Sample calculation to determine energy recovery (%) and ECR.

Author Contributions: Conceptualization, A.Y., R.V., and S.R.C.; methodology and validation, A.Y., A.K.S.V., and S.S.; formal analysis, investigation, data curation, writing—original draft preparation, and writing—review and editing, A.Y. and R.V.; resources, supervision, project administration, and funding acquisition, R.V. and S.R.C. All authors have read and agreed to the published version of the manuscript.

Funding: This research was funded by the DEPARTMENT OF SCIENCE AND TECHNOLOGY (DST), INDIA, in the form of a grant to National Center for Combustion Research and Development at Indian Institute of Technology Madras.

Conflicts of Interest: The authors declare no conflict of interest.

References

1. McLaughlin, O.; Mawhood, B.; Jamieson, C.; Slade, R. Rice straw for bioenergy: The effectiveness of policymaking and implementation in Asia. In Proceedings of the 24th European Biomass Conference and Exhibition, Amsterdam, The Netherlands, 6–9 June 2016; pp. 1540–1554. [CrossRef]
2. Sarma, S.D. Paddy Residue Burning: Drivers, Challenges and Potential Solutions. Available online: http://www.teriin.org/article/paddy-residue-burning-drivers-challenges-and-potential-solutions (accessed on 20 March 2020).
3. Nanda, S.; Mohammad, J.; Reddy, S.N.; Kozinski, J.A.; Dalai, A.K. Pathways of lignocellulosic biomass conversion to renewable fuels. *Biomass. Conv. Bioref.* **2014**, *4*, 157–191. [CrossRef]
4. Zhou, X.; Broadbelt, L.J.; Vinu, R. Mechanistic understanding of thermochemical conversion of polymers and lignocellulosic biomass. In *Advances in Chemical Engineering: Thermochemical Process Engineering*; Van Geem, K., Ed.; Elsevier: Amsterdam, The Netherlands, 2016; Volume 49, pp. 95–198.
5. Tekin, K.; Karagöz, S.; Bektaş, S. A review of hydrothermal biomass processing. *Renew. Sustain. Energy Rev.* **2014**, *40*, 673–687. [CrossRef]
6. Peterson, A.A.; Vogel, F.; Lachance, R.P.; Froling, M.; Antal, J.M.J.; Tester, J.W. Thermochemical biofuel production in hydrothermal media: A review of sub and supercritical water technologies. *Energy Environ. Sci.* **2008**, *1*, 32–65. [CrossRef]
7. Toor, S.S.; Rosendahl, L.; Rudolf, A. Hydrothermal liquefaction of biomass: A review of subcritical water technologies. *Energy* **2011**, *36*, 2328–2342. [CrossRef]
8. Liu, Z.; Zhang, F.S. Effects of various solvents on the liquefaction of biomass to produce fuels and chemical feedstocks. *Energy Convers. Manag.* **2008**, *49*, 3498–3504. [CrossRef]
9. Wang, Y.; Wang, H.; Lin, H.; Zheng, Y.; Zhao, J.; Pelletier, A.; Li, K. Effects of solvents and catalysts in liquefaction of pinewood sawdust for the production of bio-crudes. *Biomass Bioenergy* **2013**, *59*, 158–167. [CrossRef]
10. Zhu, W.W.; Zong, Z.M.; Yan, H.L.; Zhao, Y.P.; Lu, Y.; Wei, X.Y.; Zhang, D. Cornstalk liquefaction in methanol/water mixed solvent. *Fuel Process. Technol.* **2014**, *117*, 1–7. [CrossRef]
11. Pińkowska, H.; Wolak, P.; Złocińska, A. Hydrothermal decomposition of alkali lignin in sub- and supercritical water. *Chem. Eng. J.* **2012**, *187*, 410–414. [CrossRef]
12. Yuan, Z.; Cheng, S.; Leitch, M.; Xu, C. Hydrolytic degradation of alkaline lignin in hot-compressed water and ethanol. *Bioresour. Technol.* **2010**, *101*, 9308–9313. [CrossRef]
13. Malins, K.; Kampars, V.; Brinks, J.; Neibolte, I.; Murnieks, R.; Kampare, R. Bio-crude from thermo-chemical hydro-liquefaction of wet sewage sludge. *Bioresour. Technol.* **2015**, *187*, 23–29. [CrossRef]
14. Zhang, B.; Keitz, M.; Valentas, K. Thermal effects on hydrothermal biomass liquefaction. *Appl. Biochem. Biotechnol.* **2008**, *147*, 143–150. [CrossRef] [PubMed]
15. Wei, N.; Via, B.K.; Wang, Y.; McDonald, T.; Auad, M.L. Liquefaction and substitution of swichgrass (*Panicum virgatum*) based bio-crude into epoxy resins. *Ind. Crop. Prod.* **2014**, *57*, 116–123. [CrossRef]

16. Valdez, P.J.; Savage, P.E. A reaction network for the hydrothermal liquefaction of Nannochloropsis sp. *Algal. Res.* **2013**, *2*, 416–425. [CrossRef]
17. Valdez, P.J.; Tocco, V.J.; Savage, P.E. A general kinetic model for the hydrothermal liquefaction of microalgae. *Bioresour. Technol.* **2014**, *163*, 123–127. [CrossRef]
18. Yuan, X.Z.; Li, H.; Zeng, G.M.; Tong, J.Y.; Xie, W. Sub- and supercritical liquefaction of rice straw in the presence of ethanol-water and 2-propanol-water mixture. *Energy* **2007**, *32*, 2081–2088. [CrossRef]
19. Li, H.; Yuan, X.; Zeng, G.; Tong, J.; Yan, Y.; Cao, H.; Wang, L.; Cheng, M.; Zhang, J.; Yang, D. Liquefaction of rice straw in sub- and supercritical 1,4-dioxane-water mixture. *Fuel Process. Technol.* **2009**, *90*, 657–663. [CrossRef]
20. Singh, R.; Chaudhary, K.; Biswas, B.; Balagurumurthy, B.; Bhaskar, T. Hydrothermal liquefaction of rice straw: Effect of reaction environment. *J. Supercrit. Fluids* **2015**, *104*, 70–75. [CrossRef]
21. Zhou, C.; Zhu, X.; Qian, F.; Shen, W.; Xu, H.; Zhang, S.; Chen, J. Catalytic hydrothermal liquefaction of rice straw in water/ethanol mixtures for high yields of monomeric phenols using reductive CuZnAl catalyst. *Fuel Process. Technol.* **2016**, *154*, 1–6. [CrossRef]
22. Cao, L.; Zhang, C.; Hao, S.; Luo, G.; Zhang, S.; Chen, J. Effect of glycerol as co-solvent on yields of bio-crude from rice straw through hydrothermal liquefaction. *Bioresour. Technol.* **2016**, *220*, 471–478. [CrossRef]
23. Castello, D.; Pedersen, T.H.; Rosendahl, L.A. Continuous hydrothermal liquefaction of biomass: A critical review. *Energies* **2018**, *11*, 3165. [CrossRef]
24. Brand, S.; Kim, J. Liquefaction of major lignocellulosic biomass constituents in supercritical ethanol. *Energy* **2015**, *80*, 64–74. [CrossRef]
25. Saraswat, V.K.; Bansal, R. India's Leapfrog to Methanol Economy. Available online: http://niti.gov.in/writereaddata/files/document_publication/Article%20on%20Methanol%20Economy_Website.pdf (accessed on 20 March 2020).
26. ASTM E1131-08. Standard Test Method for Compositional Analysis by Thermogravimetry. Available online: http://www.astm.org/Standards/E1131.htm (accessed on 20 March 2020).
27. Vassilev, S.V.; Baxter, D.; Andersen, L.K.; Vassileva, C.G. An overview of the chemical composition of biomass. *Fuel* **2010**, *89*, 913–933. [CrossRef]
28. Chopra, J.; Mahesh, D.; Yerrayya, A.; Vinu, R.; Rajnish, K.; Sen, R. Performance enhancement of hydrothermal liquefaction for strategic and sustainable valorization of de-oiled yeast biomass into green bio-crude. *J. Clean. Prod.* **2019**, *227*, 292–301. [CrossRef]
29. Run-dong, L.; Bing-shuo, L.; Tian-hua, Y.; Ying-hui, X. Liquefaction of rice stalk in sub- and supercritical ethanol. *J. Fuel Chem. Technol.* **2013**, *41*, 1459–1465.
30. Francis, H.E.; Lloyd, W.G. Predicting heating value from elemental composition. *J. Coal Qual.* **1983**, *2*, 2.
31. Nam, H.; Capareda, S.C.; Ashwath, N.; Kongkasawan, J. Experimental investigation of pyrolysis of rice straw using bench-scale auger, batch and fluidized bed reactors. *Energy* **2015**, *93*, 2384–2394. [CrossRef]
32. Tsai, W.T.; Lee, M.K.; Chang, Y.M. Fast pyrolysis of rice straw, sugarcane bagasse and coconut shell in an induction-heating reactor. *J. Anal. Appl. Pyrol.* **2006**, *76*, 230–237. [CrossRef]
33. Akhtar, J.; Amin, N.A.S. A review on process conditions for optimum bio-oil yield in hydrothermal liquefaction of biomass. *Renew. Sustain. Energy Rev.* **2011**, *15*, 1615–1624. [CrossRef]
34. Pedersen, T.H.; Jasiūnas, L.; Casamassima, L.; Singh, S.; Jensen, T.; Rosendahl, L.A. Synergetic hydrothermal co-liquefaction of crude glycerol and aspen wood. *Energy Convers. Manag.* **2015**, *106*, 886–889. [CrossRef]
35. Feng, S.; Wei, R.; Leitch, M.; Xu, C.C. Comparative study on lignocellulose liquefaction in water, ethanol, and water/ethanol mixture: Roles of ethanol and water. *Energy* **2018**, *155*, 234–241. [CrossRef]
36. Panagiotopoulou, P.; Vlachos, D.G. Liquid phase catalytic transfer hydrogenation of furfural over a Ru/C catalyst. *Appl. Catal. A Gen.* **2014**, *480*, 17–24. [CrossRef]
37. Li, H.; Yuan, X.; Zeng, G.; Huang, D.; Huang, H.; Tong, J.; You, Q.; Zhang, J.; Zhou, M. The formation of bio-oil from sludge by deoxy-liquefaction in supercritical ethanol. *Bioresour. Technol.* **2010**, *101*, 2860–2866. [CrossRef] [PubMed]
38. Sing, S.K.; Ekhe, J.D. Towards effective lignin conversion: HZSM-5 catalysed one-pot solvolytic depolymerisation/ hydrodeoxygenation of lignin into value added compounds. *RSC Adv.* **2014**, *4*, 27971–27978. [CrossRef]
39. Tekin, K.; Karagöz, S. Non-catalytic and catalytic hydrothermal liquefaction of biomass. *Res. Chem. Intermed.* **2013**, *39*, 485–498. [CrossRef]

40. Minowa, T.; Kondo, T.; Sudirjo, S.T. Thermochemical liquefaction of Indonesian biomass residues. *Biomass Bioenergy* **1998**, *14*, 517–524. [CrossRef]
41. Zhu, Z.; Toor, S.S.; Rosendahl, L.; Yu, D.; Chen, G. Influence of alkali catalyst on product yield and properties via hydrothermal liquefaction of barley straw. *Energy* **2015**, *80*, 284–292. [CrossRef]
42. Singh, R.; Prakash, A.; Dhiman, S.K.; Balagurumurthy, B.; Arora, A.K.; Puri, S.K.; Bhaskar, T. Hydrothermal conversion of lignin to substituted phenols and aromatic ethers. *Bioresour. Technol.* **2014**, *165*, 319–322. [CrossRef]
43. Pelzer, A.W.; Sturgeon, M.R.; Yanez, A.J.; Chupka, G.; O'Brien, M.H.; Katahira, R.; Cortright, R.D.; Woods, L.; Beckham, G.T.; Broadbelt, L.J. Acidolysis of α-O-4 aryl-ether bonds in lignin model compounds: A modeling and experimental study. *ACS Sustain. Chem. Eng.* **2015**, *3*, 1339–1347. [CrossRef]
44. Liu, A.; Park, Y.; Huang, Z.L.; Wang, B.W.; Ankumah, R.O.; Biswas, P.K. Product identification and distribution from hydrothermal conversion of walnut shells. *Energy Fuels* **2006**, *20*, 446–454. [CrossRef]
45. Cheng, S.; D'cruz, I.; Wang, M.C.; Leitch, M.; Xu, C.B. Highly efficient liquefaction of woody biomass in hot-compressed alcohol-water co-solvents. *Energy Fuels* **2010**, *24*, 4659–4667. [CrossRef]
46. Yokoyama, S.; Suzuki, A.; Murakami, M.; Ogi, T.; Koguchi, K.; Nakamura, E. Liquid fuel production from sewage sludge by catalytic conversion using sodium carbonate. *Fuel* **1987**, *66*, 1150–1155. [CrossRef]
47. Vardon, D.R.; Sharma, B.K.; Grant, V.B.; Rajagopalan, K. Thermochemical conversion of raw and defatted algal biomass via hydrothermal liquefaction and slow pyrolysis. *Bioresour. Technol.* **2012**, *109*, 178–187. [CrossRef] [PubMed]

© 2020 by the authors. Licensee MDPI, Basel, Switzerland. This article is an open access article distributed under the terms and conditions of the Creative Commons Attribution (CC BY) license (http://creativecommons.org/licenses/by/4.0/).

Article

Torrefaction of Straw from Oats and Maize for Use as a Fuel and Additive to Organic Fertilizers—TGA Analysis, Kinetics as Products for Agricultural Purposes

Szymon Szufa [1,*], Grzegorz Wielgosiński [1], Piotr Piersa [1], Justyna Czerwińska [1], Maria Dzikuć [2], Łukasz Adrian [3], Wiktoria Lewandowska [4] and Marta Marczak [5]

1. Faculty of Process and Environmental Engineering, Lodz University of Technology, Wolczanska 213, 90-924 Lodz, Poland
2. Faculty of Economics and Management, University of Zielona Góra, ul. Licealna 9, 65-246 Zielona Góra, Poland
3. Faculty of Biology and Environmental Science, University of Kardynal Stefan Wyszyński, Dewajtis 5, 01-815 Warszawa, Poland
4. Chemical Faculty, University of Lodz, Tamka 53, 91-403 Lodz, Poland
5. Faculty of Energy and Fuels, AGH University of Science and Technology, al. Mickiewicza 30, 30-059 Krakow, Poland
* Correspondence: szymon.szufa@p.lodz.pl; Tel.: +48-606-134-239

Received: 14 February 2020; Accepted: 11 April 2020; Published: 21 April 2020

Abstract: This publication presents research work which contains the optimum parameters of the agri-biomass: maize and oat straws torrefaction process. Parameters which are the most important for the torrefaction process and its products are temperature and residence time. Thermogravimetric analysis was performed as well as the torrefaction process using an electrical furnace on a laboratory scale at a temperature between 250–525 °C. These biomass torrefaction process parameters—residence time and temperature—were necessary to perform the design and construction of semi-pilot scale biomass torrefaction installations with a regimental dryer and a woody and agri-biomass regimental torrefaction reactor to perform a continuous torrefaction process using superheated steam. In the design installation the authors also focused on biochar, a bi-product of biofuel which will be used as an additive for natural bio-fertilizers. Kinetic analysis of torrefaction process using maize and oat straws was performed using NETZSCH Neo Kinetics software. It was found that kinetic analysis methods conducted with multiple heating rate experiments were much more efficient than the use of a single heating rate. The best representations of the experimental data for the straw from maize straw were found for the n-order reaction model. A thermogravimetric analysis, TG-MS analysis and VOC analysis combined with electrical furnace installation were performed on the maize and oat straw torrefaction process. The new approach in the work presented is different from that of current scientific achievements due to the fact that until now researchers have worked on performing processes on oat and maize straws by means of the torrefaction process for the production of a biochar as an additive for natural bio-fertilizers. None of them looked for economically reasonable mass loss ratios. In this work the authors made the assumption that a mass loss in the area of 45–50% is the most reasonable loss for the two mentioned agri-biomass processes. On this basis, a semi-pilot installation could be produced in a further BIOCARBON project step. The kinetic parameters which were calculated will be used to estimate the size of the apparatuses, the biomass dryer, and biomass torrefaction reactor.

Keywords: torrefaction; oats; maize; straw; biochar

1. Introduction

Worldwide, big agriculture, agri-food processing and individual farmers produce a very large amount of agri-waste each year. Taking into account straw, husks plus other various types of husk and shell, and also biomass from energy crops, it accounts for 50 billion metric tons per year. The mass of straw collected in Poland is estimated at 25–33 million Mg per year [1]. A huge amount of oats and maize waste is produced annually, as maize is one of the most common staple crops worldwide. Little research [2–4] has been carried out on the handling or reuse of oats and maize waste as biomass. The latter two biomasses are converted through processes such as drying and roasting to homogenize their physicochemical properties. In this research, the torrefaction process of the most available and quite low-cost effective feedstock, oats and maize waste, was carried out in a thermogravimetric analyzer and laboratory scale furnace in the presence of nitrogen, argon, and CO_2 gas. Other methods to treat oat and maize straw are conversion to bioethanol and to a natural fertilizer using bio-treatment methods [5]. Nevertheless, those conversion techniques generate a large amount of waste [6]. In effect the soil is acidic, poor, and usually has a yellowish color grubbing up and clearing the forest quickly: most cultivated plants quickly deteriorate on it [7]. Biochar as an additive for fertilizer has a high utility potential in agriculture [8], as a base for natural dedicated fertilizers for horticulture and greenhouse cultivation and for large-scale agriculture. In addition to litter and fodder in industrial breeding of poultry and animals, it is a liquid manure absorber and dynamic of manure composter. Its use will respond to market needs related to improving soil properties, limiting the use of mineral fertilizers and plant protection chemicals and carbon sequestration in soil. The effects of soil biochar application include PH regulation, improvement of soil water properties and soil thermal properties, better use of fertilizers by plants, reduction of water pollution, positive impact on growth and yield of plants, impact on increasing activity of microorganisms, better soil (process improvement soil-forming), long-lasting effects, carbon dioxide (CO_2) sequestration, impact on mitigation of climate change, increase of the organic content in soil, increase of soil moisture, and impact on increasing the mass and quality of the system root. The application of biochar in 2014 contributed to an increase in root colonization by mycorrhizal fungi; vegetable growth; and yield of apple, peach, and nectarine trees in 2015. The application of biochar also contributed to an increase in the water status of soil and plants [7]. Biochar, due to its properties, is also an excellent structuring additive for composting: it improves the air and water conditions of the process [9]. It significantly reduces ammonia emissions, thus reducing nitrogen losses and improving the quality of the fertilizer received. It reduces gas emissions. Using biochar in soil aids retention, thermal properties, and gas. Biochar, thanks to its ionic capacity, promotes better use of fertilizers by plants and reduces the amount of pollution outflow [10]. The global biocoal market size was estimated at 283.2 Gg in 2015. The worldwide increase of consumption of bio-organic food and the capacity of biochar to increase soil fertility and plant growth is expected to be a most important factor moving market growth. The global market is composed of regimented and unregimented producers of biocoal. In China, Brazil, Japan, and Mexico large amounts of production come from the unregimented sector. The global biocoal market is expected to reach USD 3.14 billion by 2025. According to the European Biochar Foundation, the volume of biocoal production in Europe at the end of 2014 was over 9000 Mg per year. This is constantly increasing, which, taking into consideration the slow pyrolysis (30% output of biocoal from feed), gives an annual volume of biomass of 30,000 Mg, and with fast pyrolysis up to approx. 100,000 Mg per year. The European biocoal market is predicted to increase its revenue from \$193 million to \$875 million in the period 2017 to 2025, growing at an estimated compound annual growth rate (CAGR) of 17.70%. Currently, the entire production of certified biocoal in Europe is carried out in seven industrial lines located in four countries: Germany, Austria, Switzerland, and the United Kingdom. It is estimated that the volume of biocoal production in Europe is even higher due to the large number of local, small installations (perhaps more than 100). The development of the biochar market in Poland will be driven by the demand not only for biocoal as a product but primarily for the products manufactured by its use [11]. Straw from maize and oats are used in Poland for fertilizing fields. After deduction of the needs for litter and feed, as well as the amount needed for plowing, there are overflows for different use. The volume of extra production of straw according to different sources is approx. 8–13 million Mg annually [12,13]. One solution for this situation is to use straw overflows in the energy and power sector. The heating value of straw ranges from 14.3 to 15.2 MJ/kg. In terms of energy, 1.5 Mg of straw are equivalent to about one Mg of coal [14–16]. The mass of straw that can be combusted as a solid biofuel in the energy sector is equal to a calorific value of approx. 14 million Mg of coal, which is 10% of annual coal mining production in Poland. The limit in the widespread use of straw in the energy and power sector is its dispersed diversification of physicochemical properties [1,17–20]. Another very important factor is that straw is a volumetric material, which has an impact on logistic costs such as transport and storage [21–23]. The implementation of the European Commission's policy on sustainable development, the reduction of natural resources, the development and application of waste-efficient technologies, and the development of new biodegradable and environmentally friendly products is driving the growth of this market.

2. Materials and Methods

2.1. Methods

The experimental work focused on two types of agri-biomass: straw from oats and straw from maize. Thermogravimetric analysis (TGA) was used to estimate the weight loss kinetics of carbonized maize and oat straw before the design of a semi-scale set-up where a continuous working dryer and biomass torrefaction reactor. In this work, a TG 209 Tarsus Netzsch TGA was used to perform the torrefaction process. Installation with a regimental biomass dryer and regimental reactor feed with steam will be designed in the next stage to achieve the optimal mass loss to energy loss ratio (to obtain the highest energy density). The biomass torrefaction process was done using a specifically designed set-up with a TGA analyzer for torrefaction process in inert nitrogen atmosphere—Figure 1.

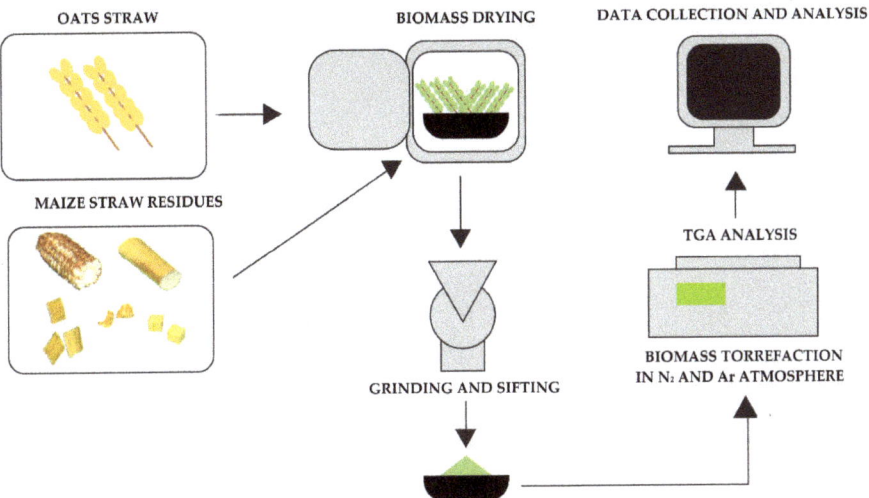

Figure 1. Principle of oats and maize straw torrefaction process with process conditions temperature and optimal mass loss using superheated steam for production of biofuel, activated carbon, and additive for natural fertilizers.

Agri-biomasses were subjected to analysis in accordance with ISO standard [24–27]. A proximate and ultimate analysis were also done after the torrefaction process. High heating values (HHVs) of the maize and oat straw after the torrefaction process were estimated in accordance with ISO standard [28]. The HHVs were determined using a bomb calorimeter (Parr Instrument Co., Model 1672, Moline, IL, USA). C, H, N, S, and O estimation analysis of the untreated biomass and torrefied samples was performed using an elemental analyzer.

2.2. Thermal Gravimetric Analysis and Kinetics

The thermogravimetric analysis (TGA) of the oats and maize straw torrefaction was done using 10 mg samples. To get an inert atmosphere, nitrogen was injected with a flow rate of 20 mL/min. The samples of oats and maize straws were placed inside the TGA analyzer furnace. The laboratory set-up with electrical furnace and TOC analyzer was calibrated with a precise (0.01 g resolution) electronic balance for the estimation of the mass drop during the process. The parameters (temperature and sample mass) were recorded using PC. For TG-MS analysis a thermogravimetric Luxx 409 PG Netzsch, coupled with a mass spectrometer QMS 403D Aeolos Netzsch were used. The tests were carried out in an argon atmosphere at 25 mL/min. Due to the need to stabilize the weight with a water jacket, the initial temperature of the 40 °C content was set. The weight of the biomass samples for the torrefaction process using TG-MS analysis was 5.0 mg. The samples were placed in 6 mm diameter Al_2O_3 crucibles.

The kinetics were determined using special NETZSCH Kinetics Neo software. Software allowed us to obtain reaction kinetics or chemical kinetics. This measures the rates of chemical processes and helps to determine reaction rates. It also takes the factors that control these rates into consideration. Information related to those points can give deep insights into the detailed molecular mechanisms behind elementary reactions. The above software is used to analyze chemical processes. NETZSCH Kinetics Neo software allows for the analysis of the biomass torrefaction process and other temperature-dependent processes. One result of such an analysis is a kinetics model which correctly corresponds to the nature of experimental data under varied temperature conditions [29]. Use of the kinetic model helps to predict a chemical system's behavior under user-defined temperature conditions [30]. Alternatively, this kind of model could be used for process optimization.

2.3. Composition of Total Organic Carbon

The value of the pollutant emission index (VOC) W_z (in mg of the pollutant tested per 1 g of fuel burned) was calculated on the basis of the relationship

$$w_z = \frac{Q \cdot c_{avr} \cdot \tau}{m_p} \quad (1)$$

where: Q —air flow rate (m^3/s),
m_p—sample mass (mg),
τ—sample torrefaction time (s),
c_{avr}—average concentration of pollutants (mg/m^3), calculated from the dependencies :

$$c_{avr} = \frac{1}{\tau} \int_0^t c(\tau) dt \quad (2)$$

3. Experimental Procedure and Device

The samples were dried so as to have homogeneous experimental conditions in an electrical oven for 4 h at 110 °C. In the experiment, each sample was processed using 2 g of sample under atmospheric pressure. In this work, a horizontal tubular reactor with a length of 600 mm and an internal diameter of 150 mm was used to perform the biomass torrefaction process, as schematically shown in Figure 2. A specific amount of biomass sample was weighed and placed in a crucible. Carbon dioxide flushing was done until the concentration of oxygen in the electrical furnace reactor was less than 1%. After flushing the reactor with CO_2 (2 L/min) the temperatures of the electrical oven chamber were increased to different temperatures set between 250 and 350 °C at a constant heating rate of 10 °C/min. After the torrefaction temperature reached the necessary experimental conditions, the heating reactor was stopped, and the inert gas was shut down. The torrefied biomass sample was then instantly removed and was weighed. The tests for the biomass torrefaction holding time were done at various holding times from 0 min to 50 min for 250, 300, 350 and 525 °C. In this experiment, the residence time in the biomass torrefaction process means the holding time after reaching a certain temperature [2,31].

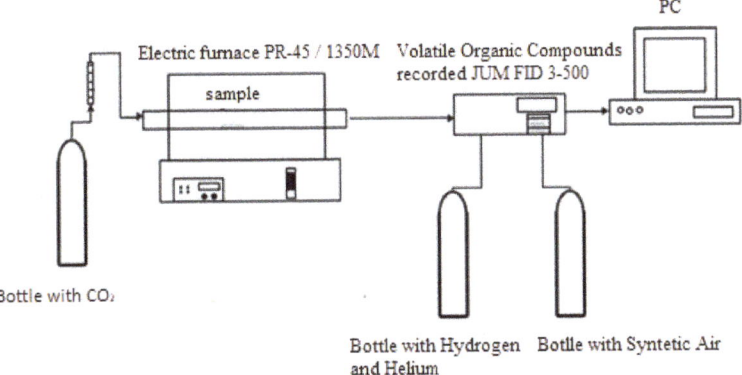

Figure 2. Installation for VOC analysis in 'torgas' during torrefaction process using electrical furnace.

The torrefaction process was conducted in a resistance electric oven PR-45/1350M in a carbon dioxide atmosphere. A previously dried sample weighing 2–2.5 g was placed on a quartz glass boat, which was put into an oven heated to the appropriate temperature for 1 h. The analysis was carried out at the following temperatures: 225, 250, 275, 300, and 525 °C. The carbon dioxide flow was 1 L/min. During this time, the emission of the sum of volatile organic compounds recorded in the JUM FID 3-500 stationary analyzer was measured. The frequency of VOC reading was 10 s. Three repetitions were performed for each temperature value. After the torrefaction process, the sample was weighed again to determine mass loss.

4. Results and Discussion

4.1. Kinetics

Kinetic analysis of the pyrolysis process of two biomass samples, maize and oats, was made based on thermogravimetric measurements using three heating rates (5, 10, and 20 K/min.). Sample weights were 10.0 ± 2 mg. For the kinetic analysis, data in a temperature range of 150–500 °C was used. Data analysis was performed using Netzsch Kinetics 3 software. Isoconversion analysis was first to be performed. This allows us to calculate the activation energy without knowing the reaction model [30]. Results of TGA analysis for the maize and oat straws are shown in Figures 3 and 4.

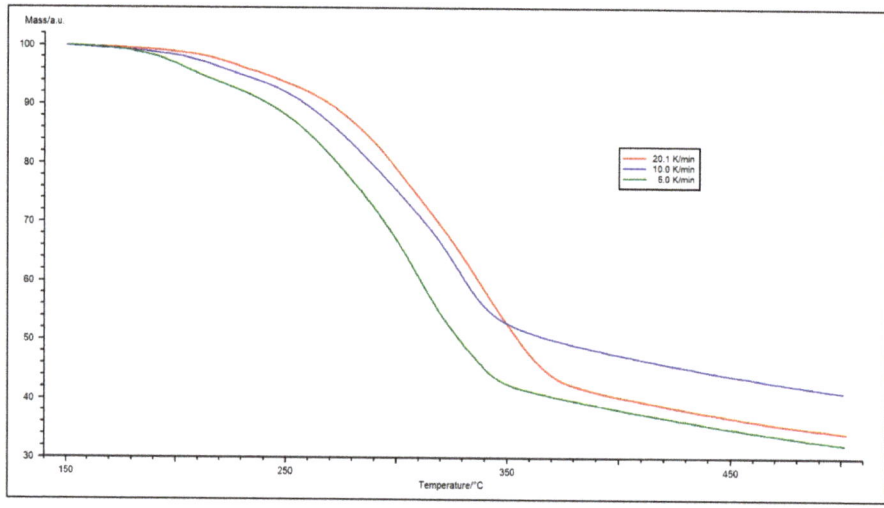

Figure 3. TG curves for three heating rates recorded for a maize straw sample.

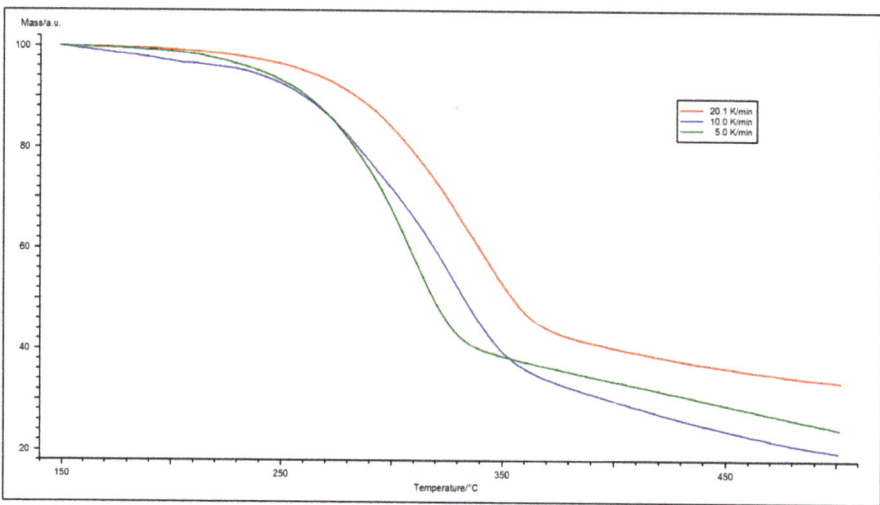

Figure 4. TG curves for three heating rates recorded for a oat straw sample.

After performing TGA analysis on maize and oat straws using 3 heating ratios, 5, 10, and 20 K/min the kinetic analysis was done using NETZSCH Kinetics Software. The first method used was a Kissinger analysis in accordance with ASTM E698 based on the assumption that the maximum (in this case the maximum of the DTG curve, i.e., the temperature at which the mass loss rate is the highest) of a one-step reaction is achieved with the same degree of conversion regardless of the heating rate. Although this assumption is only partly correct, the resulting errors are low. The Kissinger method based on Equation (1) allows the calculation of the activation energy of the thermal decomposition reaction of a solid by plotting the logarithm of the sample heating rate as a function of the inverse of temperature at the moment of the highest mass loss speed [31]

$$\ln \frac{\beta}{T_{max}^2} = \frac{-Ea}{RT_{max}} \qquad (3)$$

where: β—sample heating speed [$^K/_{min}$]. T_{max}—temperature at which the sample mass loss rate is the highest [K], A—pre-exponential coefficient [-], E_a—activation energy [$^J/_{mol}$], R—gas constant [$^J/_{mol\cdot K}$].

$$f(T) = Ae^{-\frac{E_a}{RT}} \qquad (4)$$

Reference [31], where
A—pre-exponential factor (rate constant) ($^1/_s$),
E_a—activation energy ($^J/_{mol}$),
R—gas constant ($^J/_{mol\cdot K}$),
T—temperature (K)

Friedman analysis consists of determining the logarithm of the degree of conversion and the function of temperature inverse according to Equation (5) [31]

$$\ln \frac{dx}{dt} = \ln A - \frac{E_a}{RT_{max}} + \ln f(x) \qquad (5)$$

The Ozawa–Flynn–Walla method is based on the integral form of the equation [31]

$$G(x) = \int \frac{dx}{f(x)} = \frac{A}{\beta} \int_{T_o}^{T} \exp\left(\frac{-E_a}{RT}\right) dt \qquad (6)$$

Assuming the stability $f(x)$ for a given degree of conversion, this method allows plotting the logarithm dx/dt as a function of temperature inverse and calculating activation energy (Figures 5 and 6) and pre-exponential coefficients for individual conversion levels without knowing the reaction mechanism which is presented on Figures 7 and 8.

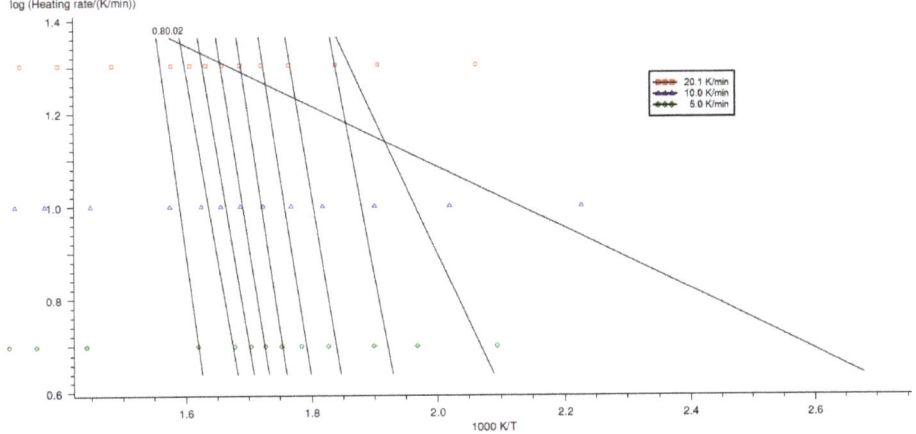

Figure 5. Graph of dx/dt logarithm as a function of temperature inverse of Ozawa–Flynn–Wall method for an oat straw sample.

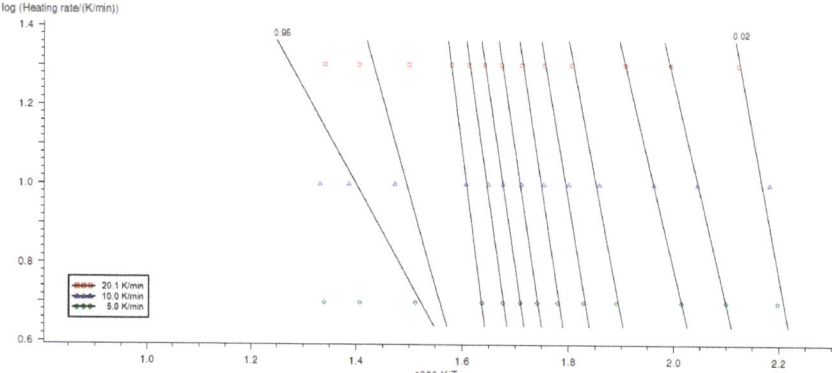

Figure 6. Graph of *dx/dt* logarithm as a function of temperature inverse of Ozawa–Flynn–Wall method for a maize straw sample.

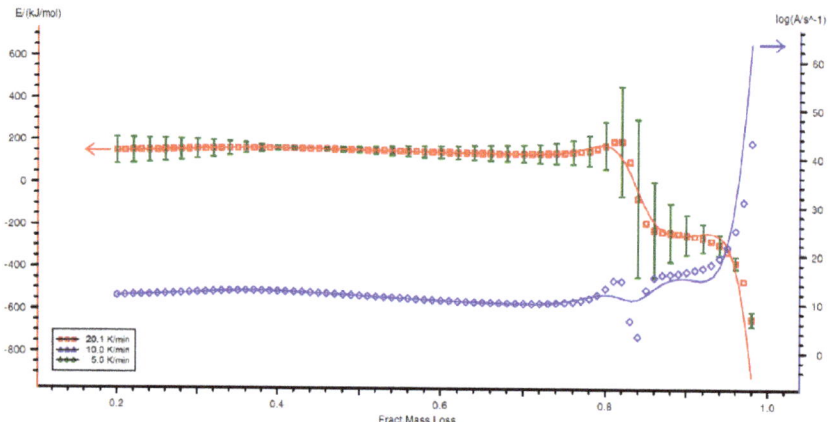

Figure 7. Activation energy and pre-exponential coefficient as a function of Ozawa–Flynn–Walla conversion for an oat straw sample.

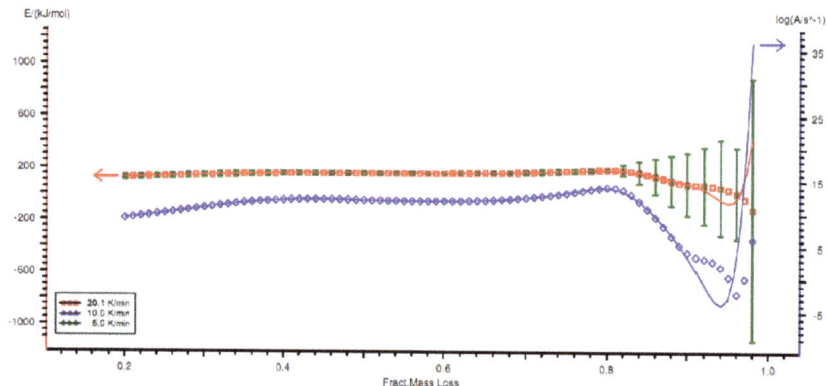

Figure 8. Activation energy and pre-exponential coefficient as a function of Ozawa–Flynn–Walla conversion for a maize straw sample.

For the analyzed biomass samples, it was assumed that the torrefaction process proceeds in one step according to the scheme

$$A - 1 \to B$$

where:

B—is the biomass after torrefaction process

Under dynamic conditions, the kinetic equation becomes a thermo-kinetic equation [31].

If

$$\beta = \frac{dT}{dt} = \text{const} > 0, \text{ then } \frac{dX}{dt} = \frac{f(T)}{\beta} f(X) \tag{7}$$

In linear form, the thermokinetic equation takes the form [31]

$$\ln \frac{g(X)}{T} = \ln \frac{A}{\beta} - \frac{E}{RT} \tag{8}$$

On the basis of the TG curves obtained as a function of temperature, using the Kinetics 3 program, an attempt was made to mathematically fit one of the models of thermal decomposition reaction. Taking into account the assumed reaction model, kinetic parameters were determined, i.e., the velocity constant (pre-exponential factor) k and activation energy E_a, as well as the fitting factor R^2 (Tables 1 and 2). In Appendix A: Table A3 summarizes the theoretical equations of the solid-state distribution models of $g(X)$, which were used to find the best fit.

Table 1. Activation energy and pre-exponential coefficient as a function of the degree of conversion by the Ozawa–Flynn–Walla method for a maize straw sample.

Fract. Mass Loss	Activation Energy (kJ/mol)	lg (A/s^{-1})
0.02	130.22 ± 44.53	−11.16
0.05	100.17 ± 1.63	5.01
0.10	98.25 ± 1.55	10.66
0.20	124.47 ± 15.81	9.34
0.30	143.37 ± 19.33	10.98
0.40	157.89 ± 18.67	12.14
0.50	159.71 ± 9.28	12.09
0.60	160.44 ± 3.21	11.98
0.70	167.82 ± 14.10	12.49
0.80	189.33 ± 7.20	14.11

Table 2. Activation energy and pre-exponential coefficient as a function of the degree of conversion by the Ozawa–Flynn–Walla method for an oat straw sample.

Fract. Mass Loss	Activation Energy (kJ/mol)	lg (A/s^{-1})
0.02	8.09 ± 41.76	−3.68
0.05	48.59 ± 73.95	1.34
0.10	130.83 ± 79.65	9.86
0.20	150.43 ± 62.49	11.53
0.30	159.74 ± 45.50	12.24
0.40	164.08 ± 11.53	12.47
0.50	156.15 ± 14.22	11.61
0.60	146.78 ± 28.35	10.66
0.70	141.47 ± 40.23	10.08
0.80	180.60 ± 111.79	13.19

The best fit for the experimental data for the oat straw sample was found for the n-order reaction model in Appendix A: Table A3 for n = 3.4235. The R^2 correlation coefficient was 0.9972, the pre-exponential coefficient 10.8466, and the activation energy 142 kJ/mol [1]. Figures 9 and 10 represent the fit of the equation with estimated coefficients fit to the experimental data.

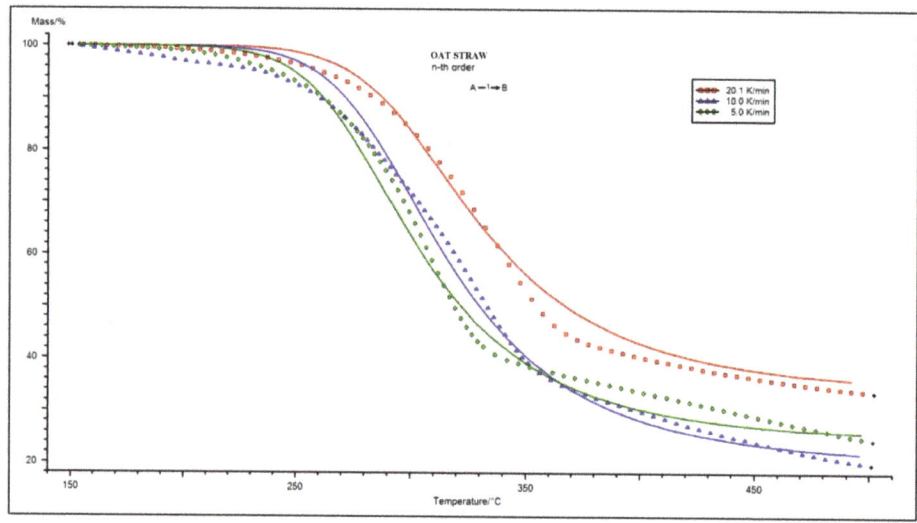

Figure 9. Fitting the calculated kinetic model to the experimental data for the oat straw sample.

The best fit for the experimental data for the maize straw sample was obtained for the n-order reaction model for n = 2.8859. The R^2 correlation coefficient was 0.9966, the pre-exponential coefficient 8.1309, and the activation energy 112 kJ/mol. Figure 9 shows the fit of the equation with calculated coefficients to the experimental data.

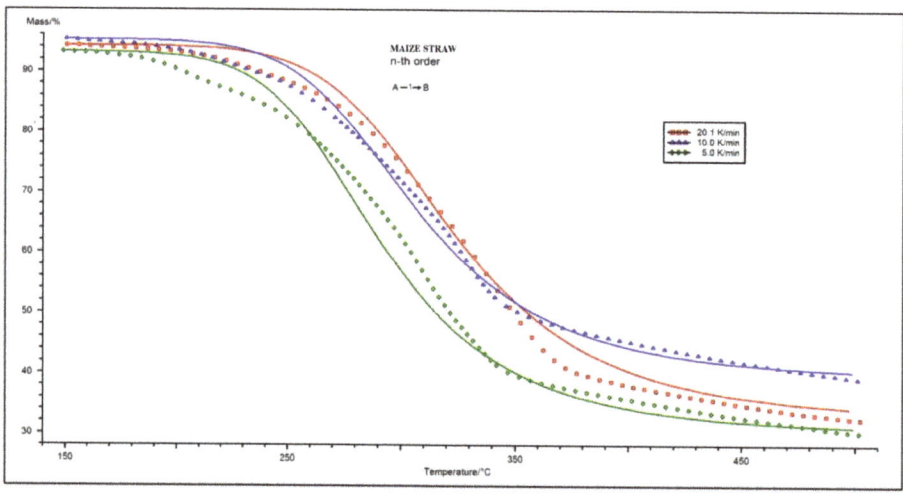

Figure 10. Fitting the calculated kinetic model to the experimental data for the maize straw sample.

It should be borne in mind that the calculated kinetic triplet is only the optimal mathematical fit of the equation to the experimental data and has no strict physical significance (reaction order) in this case. The activation energy calculated in this way is apparent: it only serves to correlate the model with the experimental data and has no meaning as in the definition [32]. It was also concluded that the methods of kinetic analysis using various heating rate results were more efficient than the method involving a single heating rate [2].

4.2. Thermogravimetric Analysis

Thermogravimetric Analysis of Oat Straw in N_2 Atmosphere Using TGA Netzsch Tarsus 209 FC

Figure 11 represents the TG curves of an oat straw sample torrefied in a nitrogen atmosphere. The lower value in the weight fraction up to 110 °C was correlated with moisture loss. The beginning of the biomass torrefaction process could be observed at temperatures of approx. 257.5 °C upon the thermal degradation of the organic components contained within the oat straw. The thermal decomposition of the oat straw was finished at a temperature of approx. 420 °C [33]. On the next increase of temperature, it was observed that there was a specific tendency for a heating rate of 20 °C/min after 400 °C. This is due to the difference of ash content caused by the inhomogeneity of the samples. Overall, it is hard to determine the proper torrefaction temperature which represents the best technical and economic feasibility levels. A decrease in mass yield results in a lower energy yield [2]. Accordingly, in this work, a kinetic analysis of the torrefaction of oat and maize straw was conducted in a temperature range which led to a conversion level of 30% for use as a carbonized solid biofuel, 50% for use as a carrier for fertilizer, and 75% for use as an activated carbon, after the removal of the moisture. It is important to obtain the kinetic parameters to get information for the proper designing of the reactor [34]. In this publication the main goal of the performed research was not to describe the fundamental chemical mechanisms for torrefaction of oats and straw. This research focuses on the calculations of apparent kinetic parameters useful for the engineering design of chemical processes and further design stage of the continuous biomass regimental dryer and biomass torrefaction regimental reactor.

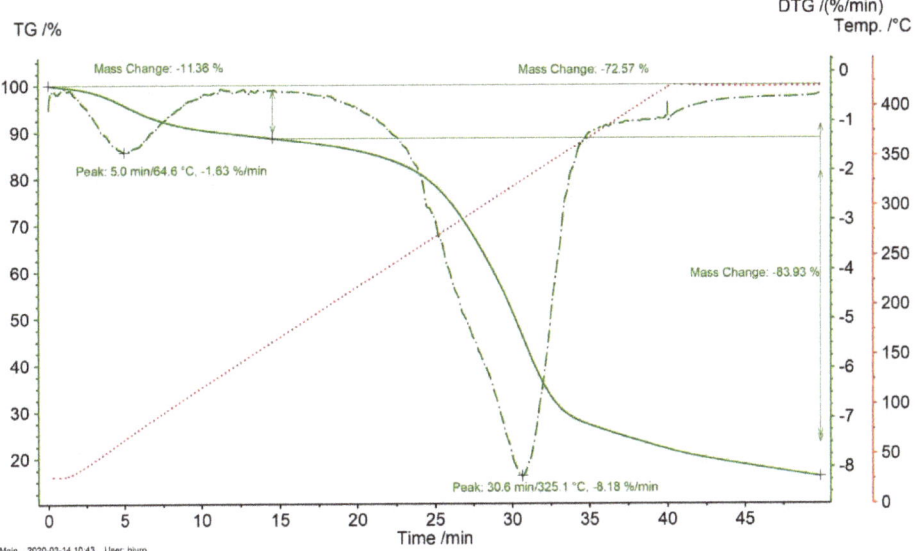

Figure 11. Thermogravimetric analysis of oat straw for the production of carbonized solid biofuel at temperature 420 °C in N_2 atmosphere.

4.3. Elemental Analysis

The results of the elemental analysis of oat and maize straw before and after the thermo-chemical conversion process are shown in Table 3. It is obvious that the weight percentage of C in the biomass torrefaction process products rose with an enhancement in the straw torrefaction process temperature, as opposed to the weight percentages of CaH and O which showed a decreasing trend. This can be explained by the fact that dehydration occurs and de-carbonization during the agri-biomass torrefaction process. It is obvious that the emission of CO_2, CO, or H_2O will result in a decrease in the H and O contents of oats and maize straw. An increasing percentage of the carbon content is only due to a decrease in the oxygen content [35,36]. These observations prove that the torrefaction process increases the energy density of the oats and maize straw by removing oxygen. It was also found that the nitrogen content did not rise significantly while the chlorine and sulphur content did not change much with a growth in the torrefaction temperature. The rise of nitrogen can also be explained as the result of a relative increase due to a reduced level of oxygen.

Table 3. Elemental analysis and technical analysis of maize and oat straws before and after torrefaction process.

Biomass Type	Moisture, (%)	C^{ad}, (%)	N^{ad}, (%)	H^{ad}, (%)	S^{ad}, (%)	Cl, (%)	Volatile ^{ad} (%)	Ash (%)	High Heating Value, (MJ/kg)
Maize straw	7.7	42.3	0.63	5.61	0.07	0.115	82.4	4.2	16.86
Torrefied maize straw:									
(2575 °C, 10 min)	3.1	51.90	0.36	4.97	0.05	0.014	63.78	4.84	21.35
(300 °C, 7 min)	1.8	54.04	0.25	4.63	0.05	0.013	51.11	6.97	22.21
(525 °C, 5 min)	1.3	58.29	0.15	3.47	0.04	0.012	37.23	10.12	26.98
Oat straw	7.8	44.10	0.65	5.87	0.09	0.230	77.8	5.0	17.74
Torrefied oat straw:									
(2575 °C, 9 min)	3.8	53.79	0.23	5.64	0.01	0.012	64.80	5.54	21.54
(300 °C, 6 min)	2.5	55.45	0.22	5.42	0.01	0.012	49.98	7.26	22.74
(525 °C, 5 min)	2.1	58.95	0.14	4.12	0.01	0.010	36.61	10.4	27.09

^{ad} Add dry basis.

On the Van Kravelen diagram (Figure 12) the modification in the O/C and H/C ratios with an increase in the oat and maize straw torrefaction process temperature can be seen. It was found that the O/C ratio goes down with rising torrefaction temperatures above 250 °C. A reduction in the O/C ratio during the torrefaction process is specific due to the formation of volatiles rich in oxygen, such as CO, CO_2, and H_2O. It can also be observed that the O/C ratios did not significantly change after 325 °C, as both C and O were lost in similar quantities through devolatilization of hydrocarbon ingredients. Another important issue is that from the torrefaction products in a solid fuel perspective the H/C ratio reduced with rising temperatures above 250 °C, while it only changed imperceptibly after 325 °C, as in the case of the O/C ratio. It can be explained that the carbon content increased relatively compared to other elements with the decrease of oxygen at temperatures of 250–325 °C, and hydrocarbon gases containing hydrogen formed above 325 °C [37,38]. The main reason for the reduction in the O/C ratio during the torrefaction process is the formation of volatiles enriched in oxygen such as H_2O and CO_2. However, more important to the higher initial value of the O/C ratio is the possibility that the highest O/C value of maize straw and oat straw after the carbonization process might be caused by the fact that there is devolatilization of cellulose and lignin, in addition to devolatilization of hemicellulose. In the end, the O/C ratio did not change significantly since both C and O amounts decrease in comparable quantities through the devolatilization process. The differences between the O/C ratios due to the torrefaction process in isothermal condition residence time were not significant, regardless of temperature [2,10].

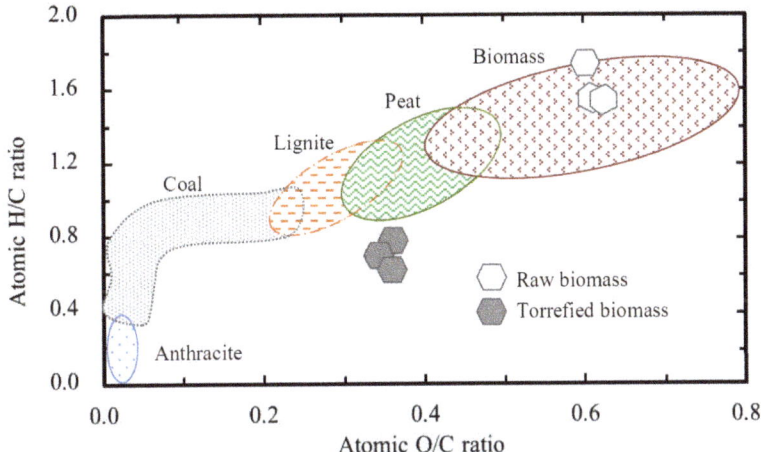

Figure 12. The Van Kravelen diagram shows location of torrefaction products: torrefied straws in diagram in relation to coal and other conventional fuels.

4.4. Mass Yield, Energy Yield, Volatile Fraction, Higher Heating Value, and Ash Content

The trajectory of the oat straw torrefaction process curves shown in Figure 11 in the case of each of the torrefaction process curves in the drawing with green points shows the end of the heating process and the beginning of holding the torrefied material at a specified temperature. As can be seen in the case of the maize straw torrefaction process carried out up to 420 °C during 10 min of holding, the sample mass reduction was 83.93%. During storage at 525 °C the mass decreased by 0.3%, as can be seen from the analysis of the trajectory of the curves presented at 525 °C. The material has already been completely degassed to illustrate the course of the process under study, and a set of TG–DTG curves is provided together with the set maximum process speeds (minima on DTG curves). Energy yield per dry raw biomass indicates the total energy preserved in the torrefied biomass. This was calculated from mass yield and higher heating values using Equation (9) and expressed as a percentage of energy content of untreated dry biomass.

$$\text{Mass Yield Y}_{mass}\ (\%) = \frac{m_{torrefied}}{m_{initial}}\ \text{dry basis} * 100\% \tag{9}$$

where: $m_{torrefied}$ = mass of biomass feedstock measured after torrefaction expressed on dry basis, $m_{initial}$ = mass of untreated (raw) biomass feedstock measured before torrefaction expressed on dry basis.

$$\text{Energy Yield Y}_{energy}\ (\%) = Y_{mass} * \frac{HHV_{torrefied\ sample}}{HHV_{raw\ sample}}\ \text{dry basis} * 100\% \tag{10}$$

where, $E_{torrefied}$ = specific energy content of biomass feedstock after torrefaction expressed on dry basis, $E_{initial}$—dry basis = specific energy content of biomass feedstock before torrefaction expressed on dry basis.

As can be seen in Figure 13, the initial evaporation of moisture from the sample is the fastest around 80 °C. In the case of torrefaction to 257 °C, the maximum process speed is 3.42 %/min at a temperature of approx. 265 °C (maximum instantaneous temperature of this process during stabilization to 257.5 °C). For maize straw torrefaction up to 300 °C, the maximum process speed is 5.99 %/min at a temperature 284.5 °C, and in the case of torrefaction up to 525 °C, the maximum process speed is 7.39 %/min at a temperature of 329 °C, with an additional minimum of 287 °C visible on the DTG curve (6.25 %/min) which indicates a two-stage distribution of the tested biomass.

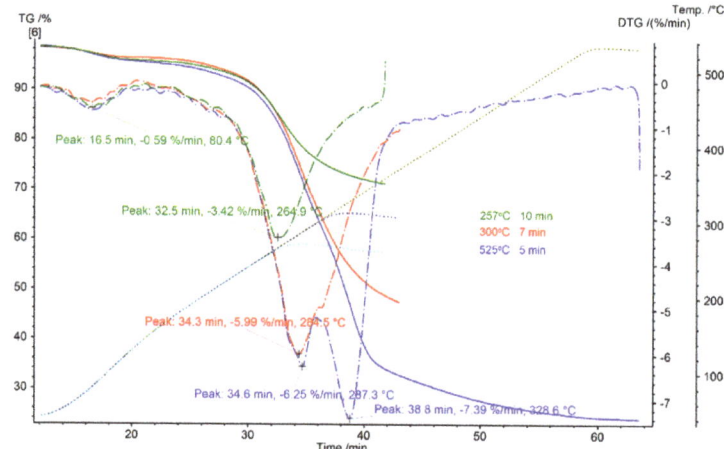

Figure 13. Summary of maize straw torrefaction curves with DTG curves.

The trajectory of torrefaction curves shown in Figure 13, in the case of each of the torrefaction process curves in the drawing with blue points, shows the end of the heating process and the beginning of holding the torrefied material at a specified temperature. As can be seen in the case of the oat straw torrefaction process carried out up to 257 °C during 10 min of holding, the sample mass reduction was 10.62% when kept for 6 min at 300 °C, the weight reduction was 11.68%, while during storage at 525 °C the mass decreased by 0.13%, and as can be seen from the analysis of the trajectory of the curves presented at 525 °C the material has already been completely degassed to illustrate the course of the process under study. A set of TG-DTG curves is provided together with a set of maximum process speeds (minima on DTG curves). As can be seen in Figure 14, the initial evaporation of moisture from the oat straw sample is the fastest around 70 °C in the case of torrefaction to 257 °C, and the maximum process speed is 3.08 %/min at a temperature of approx. 262 °C (maximum instantaneous temperature of this process during stabilization to 257.5 °C). For oat straw torrefaction up to 300 °C, the maximum process speed is 8.01%/min at a temperature around 300 °C, and in the case of torrefaction up to 525 °C, the maximum process speed is 8.81 %/min at a temperature of 305 °C.

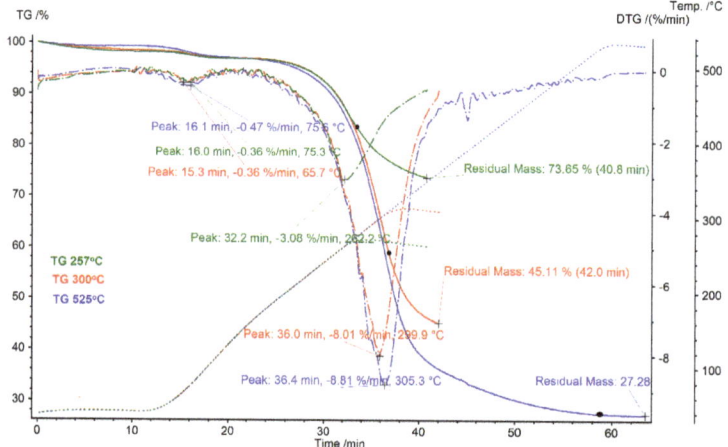

Figure 14. Summary of oat straw torrefaction curves with DTG curves.

In this paper, the torrefaction process of oat and maize straw was performed to find the optimal torrefaction temperature for different industrial applications such as fuel and as an additive for natural fertilizers and activated carbon [39,40]. The experiments with respect to torrefaction temperature were conducted at temperatures of 250–525 °C. In most of the available publications the torrefaction process is commonly performed at temperatures of around 250–300 °C, and the thermal treatment above 300 °C is referred to as pyrolysis. The HHV both for oat and maize straw of all samples rises with the increasing torrefaction temperature. A visible rise in the HHV up to 400 °C is related to the removal of oxygen and hydrogen. A rise in the temperature above 500 °C shows a reduction in the HHV. This can be explained by the fact that the pyrolysis reaction occurred at a higher temperature above 450 °C. Calculations of the energy yields are useful measures during the torrefaction process of oat and maize straws. They can be derived from the mass yield, as was done by Bridgeman et al. [11]. In this paper, the energy yield decreased slowly from 92% to 35% with a rise in the temperature of the biomass torrefaction process. The energy yield delivered in this study is based on the HHV [2]. The same procedure like for biomass torrefaction process of oat straw was was done for maize straw Figure 15a–c. The lowest mass reduction for oat straw occurs at a temperature of 257.5 °C and holding time 5 min, and it represented 19.44% of the untreated oat straw (Figure 16a). At a 257.5 °C temperature regime it was found that a holding time of 10 min represents the closest to 30% mass loss, 24.21%, which is expected for torrefaction products used as a fuel. Mass reduction of oat straw during a torrefaction process at 300 °C shows the closest to a 50% mass reduction during the experiment when the sample was held for 6 min and mass loss corresponding to this resident time was 48.98%. These are the input data that were used to calculate the profitability of using torrefied oat straw and straw from maize as additives to organic fertilizers. It is essential for the economy [40–45] of the overall torrefaction process that there is a high carbonization rate correlated with a low energy input for the process. Therefore, it was assumed that oat and maize straw use after the torrefaction process as additives for natural fertilizers should not exceed 50% of the mass reduction and the holding time to achieve it should be the lowest as possible. For the production of activated carbon, it was assumed that the mass loss should not exceed 75%, and during an oat straw torrefaction process at 525 °C it was found that the shortest residence time so as not to exceed this mass reduction was 6 min.

In Figure 16a–c the mass residua of oat straw torrefaction products are shown proceeding in three temperatures, 257.5, 300 and 525 °C, in different holding times under isothermal conditions (from 5 to 10 min) using three different heating rates, 5, 10, and 20 K/min. Thanks to the use of three different heating ratios, data were confirmed on the elemental composition and thermal stability of the isotherm research material. Thanks to analysis performed at three temperatures of 525 °C (process conditions for activated carbon), 300 °C (biocarbon for carrier of natural bio-fertilizer) and 257.5 °C (carbonized solid biofuel), it can be concluded that the mass loss ratio rises with the length of the isotherm in each case. A change can be seen in the sample at 300 °C, which represents very large variance between the values before and after biomass torrefaction process—max. (before) 52.04% and min. (after) 21.28%. Nevertheless, at a temperature of 257.5 °C the residual masses are the largest of all three temperatures. This is a very good example showing the reduction in mass with increase of temperature, and the per cent of mass of the biomass at 525 °C is the lowest of all measurements. At 257.5 and 525 °C, as the isotherm time increases, a slight increase in weight loss is noted. During extension of the isotherm at 300 °C, the mass of oat straw decreases. This research presents the preservation of straw mass from oats and maize after thermogravimetric analysis. Samples with maize straw and oats were subjected to dynamic and isothermal measurements at a 10 K/min heating rate. Thanks to that, information on thermal stability and the elemental composition of the isotherm research material were obtained, divided into five samples, from 5 to 10 min, according to the measurements found at three temperatures of 525, 300, and 257.5 °C [1].

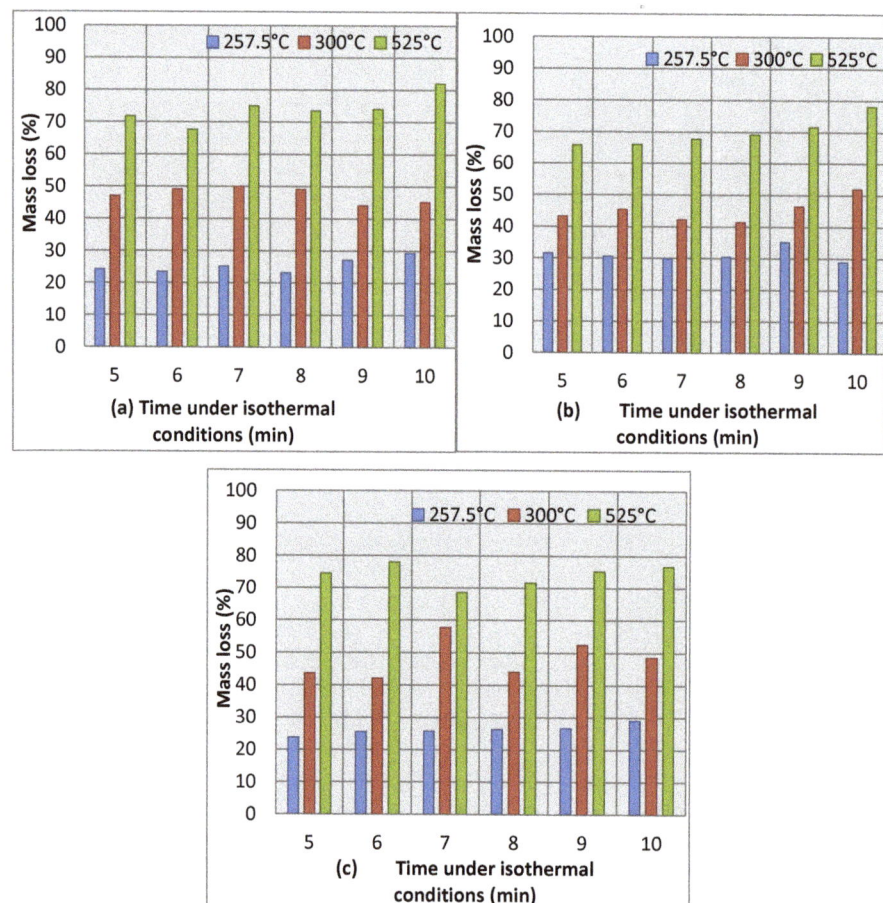

Figure 15. Maize straw torrefaction process TGA analysis results and process conditions as in temperature and optimal mass loss for production of biofuel, additive for natural fertilizers, activated carbon (Figure 15a heating rate 5 K/min, Figure 15b 10 K/min, Figure 15c 20 K/min).

This work presents a thermogravimetric analysis of oat and maize straw. We subjected biomass to dynamic and isothermal measurements. Maize straw was subjected to thermal analysis, which includes measuring the change in sample mass depending on temperature or time. The change in sample mass depends on the temperature or time dm/dt and is recorded by means of a thermogravimetric curve. TGA maize straw results were compared with oat straw TGA analysis. We divided isotherms into 5 parts with lengths of 5, 6, 7, 8, 9, and 10 min at three different heating rates, 5, 10, and 20 K/min. Our sample consisted of five maize elements: crumb, bark, leaves, grains from the cob and stigmas of corn stalks. According to the measurements found at three temperatures of 525, 300, and 257.5 °C for both types of straw, it could be found that the mass loss ratio rises with the length of the isotherm in each case. For maize straw at 300 °C, the weight loss value for the 5-min isotherm is 43.22%, and for 10 min under isothermal conditions the mass loss was 51.89%. In the case of oat straw and 300 °C temperature, this loss is slightly larger, because for 5 min under isothermal conditions the weight loss was 47.96% and for 10 min it was 78.72% Other results were obtained in the analysis of oat straw, because at a temperature of 257.5 °C the weight loss is proportional to the isothermal time and is estimated to be a difference of about 5%. During the extension of the isotherm at 300 °C, the mass of straw from maize and oats dropped sharply (in the case of oats in higher values). Therefore, we dare say that not every temperature and type of biomass lengthening of the isotope results in a sudden change in mass. Oat straw is a biomass more susceptible to the effects of low temperatures than straw from maize. We presented the TGA result in the form of a curve in which the mass is plotted depending on the temperature. To supplement the chart with more accurate data, we used the first derivative graph, which illustrates the speed of change in sample mass during the measurement. This curve clearly shows the maxima and minima that correspond to the changes in mass. In addition, we made a bar chart containing the results of all 18 measurements. This presents the weight loss at each temperature and isotherms in percentage form.

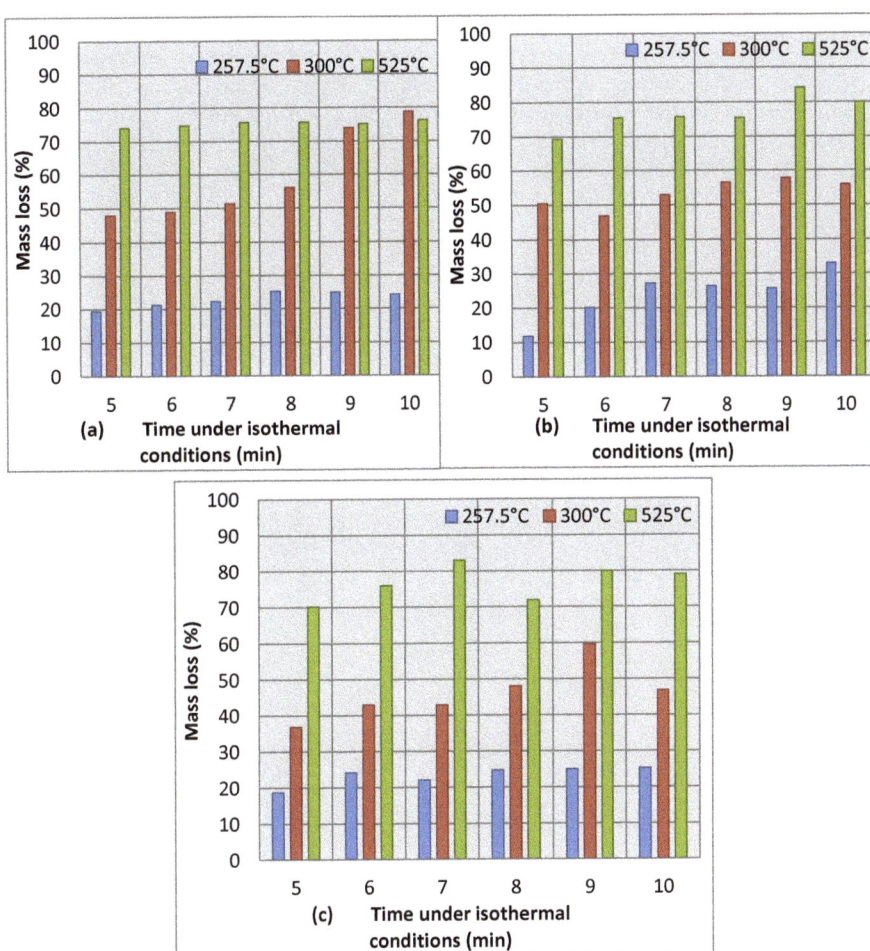

Figure 16. Oat straw torrefaction process TGA analysis results and process conditions as in temperature and optimal mass loss for production of biofuel, additive for natural fertilizers, activated carbon (Figure 16a heating rate 5 K/min, Figure 16b 10 K/min, Figure 16c 20 K/min).

4.5. Effect of the Torrefaction Residence Time

A TGA analysis of oat and maize straws found that torrefaction process temperatures above 300 °C resulted in mass yields and decreased energy. Effects of the torrefaction process residual time were studied for temperatures below 400 °C. In Figures 15 and 16a correlation between the effects of the agri-biomass torrefaction residence time on the characteristics of oats and maize straw torrefaction can be observed. It was concluded that the effect of torrefaction residence time on volatile matter and ash content was not as big as biomass torrefaction process temperature. In Figures 16 and 17, the mass yield reduces with an increase in the torrefaction residence time. This could be explained by the reduction in the water content and volatile content of the straw. Nevertheless, there was a meaningful mass loss at the beginning of the biomass torrefaction process while the change of mass yield was not so important with a longer torrefaction residence time. This can be explained by the decomposition of more reactive components at the beginning of the torrefaction [20]. It was also concluded that the mass yield reduces with the torrefaction residence time. It can be inferred from Figure 10 that the HHV rises with a longer torrefaction time. However, the biomass torrefaction residence time has been shown to be less significant than the temperature in all experiments conducted thus far [20]. The shortest torrefaction residence time can differ depending on the torrefaction temperature, biomass type, the physical and chemical properties of the biomass, and its intended use for a specific sector [46]. Arias et al. [7] concluded that there was a small improvement in biomass grind ability at 240 °C if the torrefaction residence time was longer than 30 min. In another study, Bergman et al. [21] founded that torrefaction should be performed for 17 min at 280 °C for co-firing applications [2].

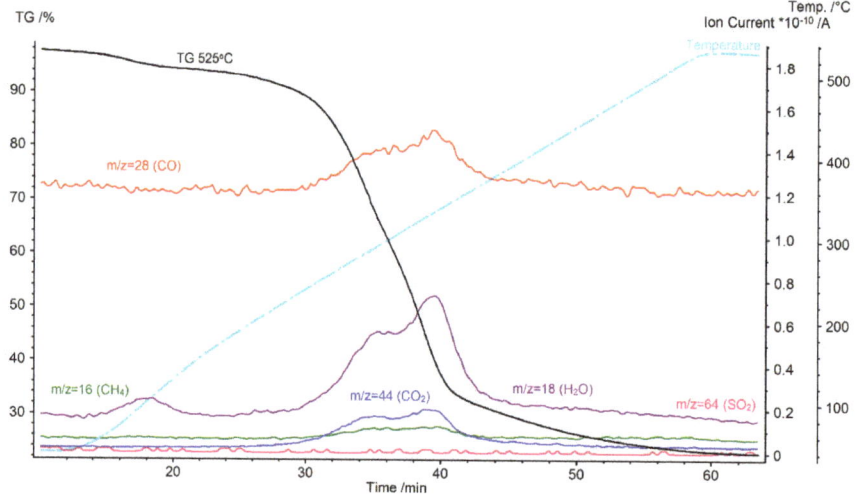

Figure 17. TG curve and ion current curves for the maize straw torrefaction process carried out up to 525 °C.

4.6. TG-MS Analysis of the Gaseous Products

For each process, tests were carried out on the evolution of gases using a mass spectrometer coupled with a thermogravimetric analyzer. TG-MS analysis was done to find out the gaseous compounds of torgas during the torrefaction process of oat and maize straw. Signals from the gas spectrometer were recorded in bar graph mode, i.e., a range from 3 to 84 [amu] was scanned. The data obtained were subjected to detailed analysis and it was found that during the implemented processes only increases in ionic currents corresponding to such gases were visible: H_2O, CH_4, CO i CO_2, and SO_2. The trajectory of ionic currents for the main ions of the above gases are shown in Figures 17 and 18 and in Appendix A: Figures A7–A10. Analysis shows that during higher torrefaction, temperature corresponds to the generation of CO, which takes place at a temperature of 300 °C for oat straw and 295 °C for maize straw.

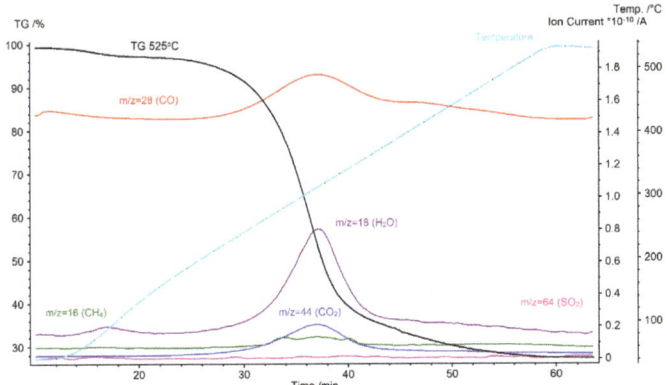

Figure 18. TG curve and ion current curves for the oat straw torrefaction process carried out up to 525 °C.

As can be seen from the analysis of the curves in Figures 17 and 18, no SO_2 secretion was observed in any of the processes studied. In addition, for processes carried out up to temperatures of 257.5 and 300 °C, one maximum emission of such gases is visible, such as CO, CO_2, CH_4, and two excretion maxima for water. The temperatures of these maxima correspond to the temperatures read from the DTG curves. In the case of a process carried out at a temperature of 525 °C there are two maxima of evolution of gases such as CO, CO_2, and CH_4 and three secretion maxima for H_2O. The temperatures of these maxima correspond to the temperatures read from DTG curves. In Figure 18, characteristic change can be found in samples at 525 °C, which shows differences between values over 12%; but in TGA straw from oats, this change took place at 300 °C and reached a difference of over 30%. Attention should be paid to a temperature of 257.5 °C, because the residual masses are the largest from all three temperatures. The percentage values of remaining mass at 525 °C are the smallest of all measurements for both biomass types [1]. At 257.5 °C for maize straw, the defect is non-standard and disproportionate (it varies within a few per cent). The reason for this may be a too low temperature test for the tested biomass.

4.7. Composition of Total Organic Carbon

The mass yield of oats and maize straw after a torrefaction process at temperatures up to 300 °C indicated a reduction in mass similar to that of the untreated straw before the torrefaction process. This showed that the extent of torrefaction for the oats and maize straw up to 300 °C was insignificant equated to these values above 300 °C. There were two main causes of the reduction in mass of the dried or torrefied products. First is moisture loss at the beginning of the process up to 110 °C, with the other being thermal decomposition forming volatile gaseous products such as H_2O, CO, CO_2, acetic acid, and other organics (Figures 18 and 19 and Figures A7–A10). The reduction in the mass yield is correlated with well known thermal effects which is shown in the loss of moisture, followed by the depolymerization of the secondary cell-wall constituents hemicellulose, cellulose, and lignin Figure 16. The decrease in mass during a torrefaction process at a lower temperature is explained as mainly caused by the loss of water content. To be accurate, torrefaction did not appear in oat and maize straw samples at temperatures below 300 °C. In a biomass torrefaction process at temperatures above 300 °C, the decrease is related to the thermal decomposition of the biomass [2]. The volatile parts show a reduction while the ash content increases with biomass torrefaction temperatures up to 525 °C. It is more important that the volatile matter content rose while the ash content reduced at temperatures above 525 °C [47]. The high heating value constantly increases with the rise of volatile materials and the decrease of ash content.

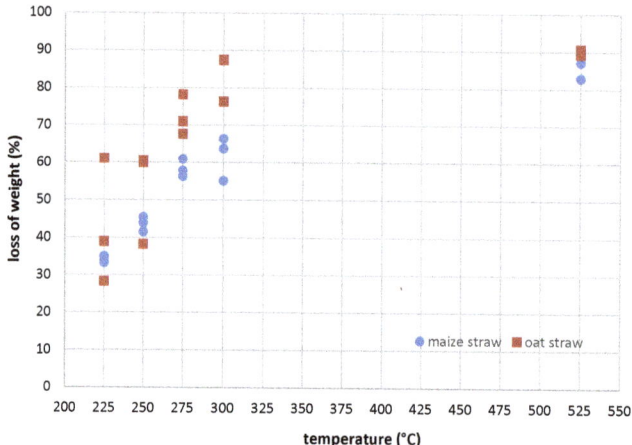

Figure 19. Mass loss of maize and oat straw after torrefaction process in CO_2 atmosphere using electrical furnace.

Figures 20 and 21 shows measurement results of VOC for maize and oat straws.

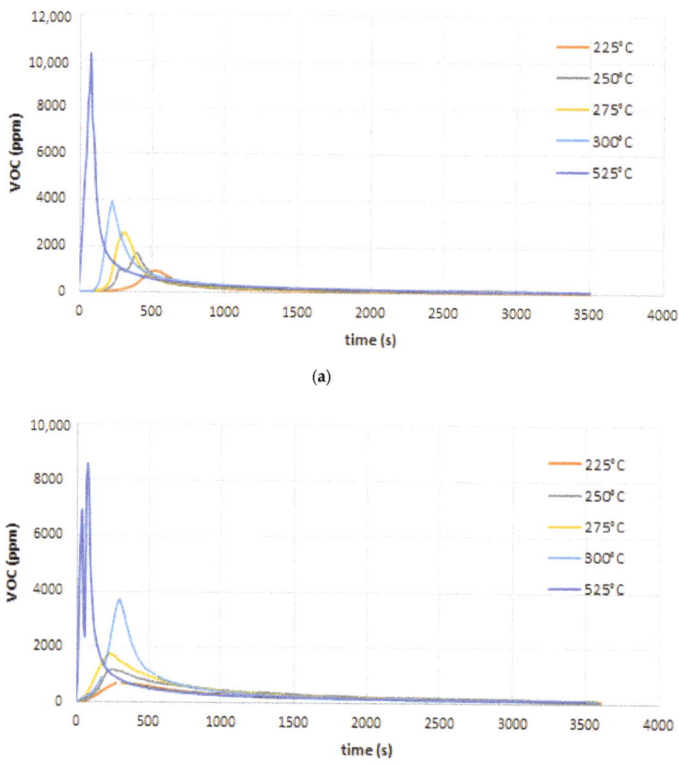

Figure 20. (a) VOC emission in ppm of maize straw during torrefaction process in electrical furnace in CO_2 atmosphere. (b) VOC emission in ppm of oat straw during torrefaction process in electrical furnace in CO_2 atmosphere.

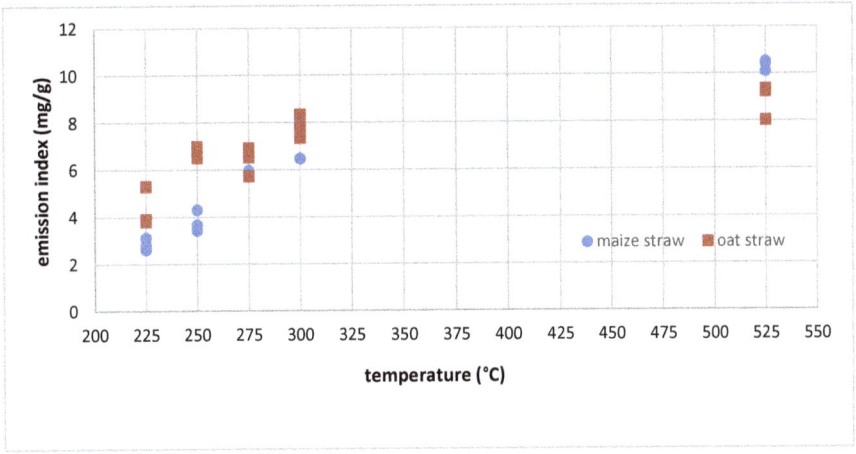

Figure 21. Emission index for maize and oat straw torrefaction process using electrical furnace.

Research results show that torrefaction of maize straw at temperatures lower than 370 °C is more suitable for the production of solid biofuels. A process temperature above 370 °C is a temperature at which hydrocarbon gas emissions begin, which has a negative impact on the energy efficiency of overall process. From Figure 14, temperatures between 290 and 310 °C were considered as the optimum torrefaction temperatures for maize straw to obtain mass loss in a range of 45–55%. Above 310 °C the HHV remained at 21 MJ/kg in this area and the energy yield remained above 70%. The tradeoff of the energy yield against the HHV limits the torrefaction region. Nevertheless, the torrefaction area is different with the initial condition of the untreated material, the heating rate, and the chemical composition [2,48].

5. Conclusions

In this paper, two kinds of agri-biomass (oat straw and maize straw) torrefaction process conditions were measured to obtain three kinds of products as follows. The authors quantified different biomass torrefaction process parameters (kinetics, optimal torrefaction process temperature, and residence time for specific mass loss ratio). An analysis of the above data shows that above 400 °C, the sample torrefaction process is completed. These results can be explained by the fact that compounds with oxygen are emitted at temperatures lower than those for hydrocarbon gases. It was concluded that biomass the torrefaction process temperature had a bigger effect on the torrefaction of maize and oat straw compared with the torrefaction residence time. From an elemental analysis, the weight percentage of C in the maize and oat straw rose with an increase in the torrefaction temperature. On the other hand, the weight percentages of H and O tended to reduce. Accordingly, the O/C and H/C ratios decreased with a rise in the torrefaction temperature. From an analysis of the gaseous products formed during biomass torrefaction, it was concluded that the compounds with oxygen were emitted at a temperature lower than that for hydrocarbon gases. From this study, it was found that temperatures between 290 and 330 °C are the most optimum torrefaction temperatures for maize and oat straw, although the actual temperature can be different depending on the initial condition of the untreated material, the heating rate, and the chemical composition of the raw biomass [2]. It was also found that kinetic analysis methods using multiple heating rate experiments are more efficient compared to the use of a single heating rate. The biomass torrefaction process proceeds with the release of volatile organic compounds. This emission is very rapid in the initial period, although it decreases over time. VOC emission is the result of anaerobic decomposition of organic matter forming biomass and is higher when increasing the temperature of the torrefaction process. Maximum VOC concentrations in torgas rises up to 10,000 ppm, while the emission factor (related to mass unit) reaches a value of several mg/g of torrefied biomass. High VOC concentrations in the torgas provide a chance for use as a source of heat (fuel) supplied to the torrefaction process.

Author Contributions: S.S. and J.C.; methodology, G.W., M.D.; software, Ł.A.; validation, S.S., M.M. and P.K.; formal analysis, S.S. and G.W.; investigation, S.S., J.C. and L.W.; resources, S.S. P.P.; data curation, S.S. and L.W.; writing—original draft preparation, S.S.; writing—review and editing, S.S. and J.C.; visualization, W.L.; supervision, S.S. and G.W.; project administration, S.S.; funding acquisition, S.S. All authors have read and agreed to the published version of the manuscript.

Funding: The studies presented were financed by the National Center of Research and Development (NCBR) Poland under the research program LIDER. The research and development project is entitled by the acronym BIOCARBON, with the title "Modern technology biomass torrefaction process to produce fuel mixtures, biocoal as additives for fertilizers, activated carbon for energy sector, agriculture, civil engineering and chemical industry", "LIDER IX" NCBR 2014-2020 (grant no. 0155/L-9/2017). The studies presented were financed by the National Center of Research and Development (NCBR) Poland under the sectorial research program. The research and development project is entitled by the acronym: DRPBIOCOAL: "Development of an innovative technology of biomass torrefaction to produce biochar, fuel and activated carbon for the needs of agriculture, energy and industry", Grant from the National Center for Research and Development (NCBiR) in Warsaw - POIR (Operation Program for Inteligent Development 1.2 "Sectoral R&D programs" 2014-2020 (PBSE) in 2017, (grant no. POIR.01.02.00-00-0243/17). This study was conducted and financed in the framework of the research project "Economic aspects of low carbon development in the countries of the Visegrad Group", (grant no. 2018/31/B/HS4/00485), granted by the National Science Centre, Poland, program OPUS.

Conflicts of Interest: The authors declare no conflict of interest.

Appendix A

Figures A1 and A2 present data analysis for a sample of oats and maize straws by the Kissinger method (ASTM E698).

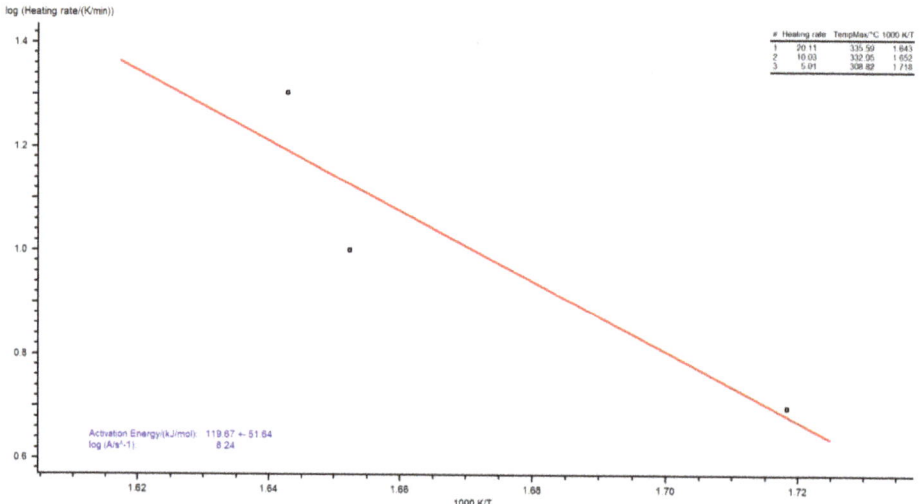

Figure A1. TG curves for three heating rates recorded for a oat straw sample.

The activation energy values obtained are 119 and 125 kJ/mol for oats and maize, respectively. Note the very high relative error of this method (51 and 34 kJ/mol for oats and maize, respectively). The degree of conversion and reaction speed were determined from the formulas [31]

$$X = \frac{m_0 - m}{m_0 - m_k} \quad (A1)$$

$$\frac{dX}{dt} = \frac{dm}{dt} * \frac{1}{m_k - m_0} \quad (A2)$$

where:
m—mass at a given time t (mg),
m_0—initial mass (mg),
m_k—mass remaining after the reaction (mg)

The reaction rate is a function of temperature (31):

$$\frac{dX}{dt} = f(X) * f(T) \quad (A3)$$

where:
t—time (s),
X—conversion level,
f(X)—reaction kinetic model,
f(T)—Arrhenius equation, given by

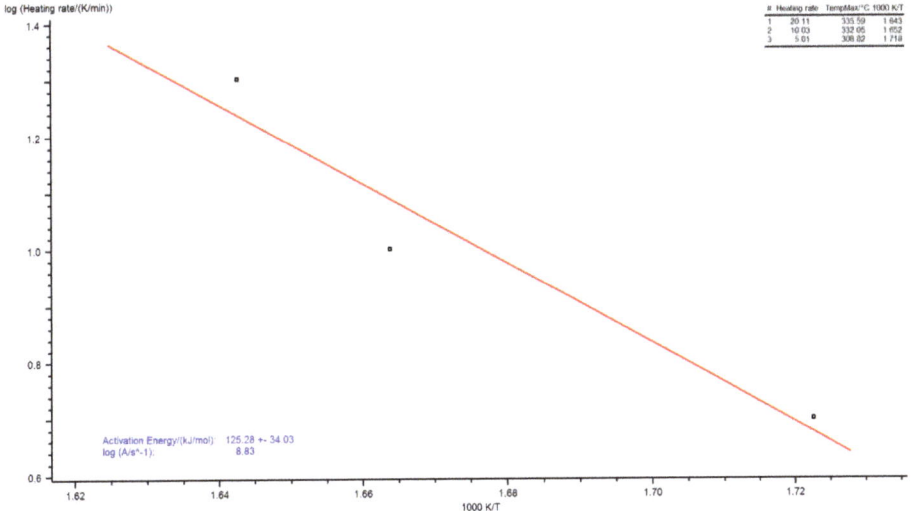

Figure A2. TG curves for three heating rates recorded for a maize straw sample.

Because f(x) is constant for a given conversion, the log with dx/dt as a function of temperature inverse gives straight lines with a slope equal to E_a/R. Figures A3 and A4 show logs of dx/dt logarithm as a function of temperature inverse determined by Friedman's method for maize and oat straw samples.

Friedman's method also makes it possible to present the value of activation energy and pre-exponential coefficient as a function of the degree of conversion. Figures A5 and A6 present these graphs for a sample of oats and maize.

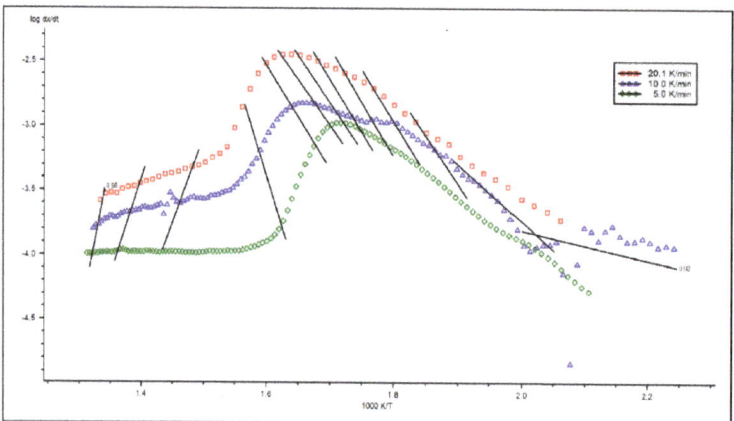

Figure A3. Log *dx/dt* graph as a function of temperature inverse using the Friedman method for an oat straw sample.

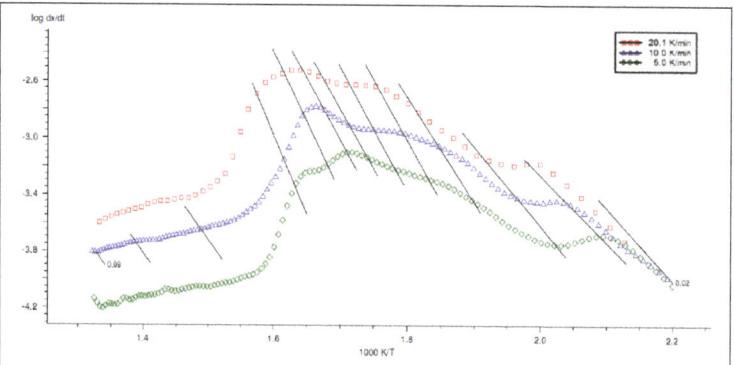

Figure A4. Graph of *dx/dt* logarithm as a function of temperature inverse using the Friedman method for a maize straw sample.

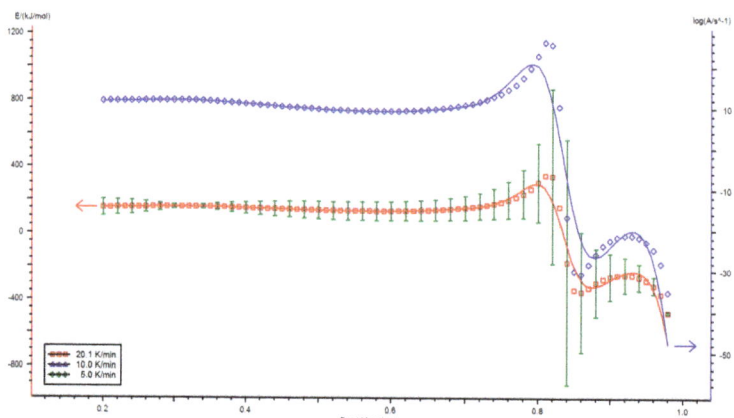

Figure A5. Activation energy and pre-exponential coefficient as a function of the degree of conversion by the Friedman method for an oat straw sample.

Analysis of Figures A5 and A6 shows that the activation energies of oat and maize samples vary depending on the extent of conversion.

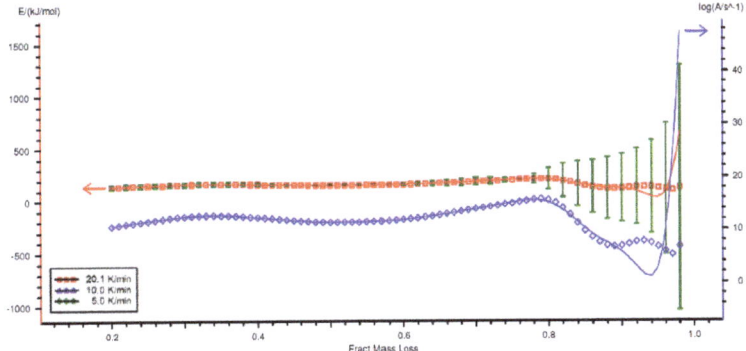

Figure A6. Activation energy and pre-exponential coefficient as a function of the degree of conversion by the Friedman method for a maize straw sample.

These results were also collected in tabular form (Tables A1–A3).

Table A1. Activation energy and pre-exponential coefficient as a function of the degree of conversion by the Friedman method for a maize straw sample.

Fract. Mass Loss	Activation Energy (kJ/mol)	lg (A/s−1)
0.02	98.35 ± 0.15	7.31
0.05	90.03 ± 6.69	6.16
0.10	107.15 ± 5.50	7.62
0.20	141.67 ± 23.30	10.70
0.30	163.62 ± 19.12	12.52
0.40	165.03 ± 12.55	12.37
0.50	157.00 ± 5.09	11.44
0.60	164.05 ± 7.43	11.95
0.70	189.61 ± 37.88	13.97
0.80	214.67 ± 99.54	15.65

The variability of the activation energy value as a function of the degree of conversion most probably indicates the fact that the pyrolysis reactions of oat and maize samples are not homogeneous and consist of many overlapping and parallel thermal decomposition processes of individual components of these biomass types.

Table A2. Activation energy and pre-exponential coefficient as a function of the degree of conversion by the Friedman method for an oat straw sample.

Fract. Mass Loss	Activation Energy (kJ/mol)	lg (A/s^{-1})
0.02	21.63 ± 51.95	−1.54
0.05	83.53 ± 61.91	5.01
0.10	141.68 ± 62.75	10.66
0.20	155.12 ± 49.02	11.71
0.30	159.65 ± 12.10	11.92
0.40	152.36 ± 34.47	11.10
0.50	138.92 ± 51.86	9.81
0.60	133.57 ± 56.59	9.26
0.70	152.72 ± 71.48	10.75
0.80	305.52 ± 234.60	22.83

Table A3. Activation energy and pre-exponential coefficient as a function of the degree of conversion by the Ozawa–Flynn–Walla method for an oat straw sample.

Model	Process	Function g(x)
D1	One-dimensional diffusion	x^2
D2	Two-dimensional, cylindrical diffusion	$(1 - x)\ln(1 - x) + x$
D3	Three-dimensional, spherical diffusion	$[1 - (1 - x)^{1/3}]^2$
D4	Three-dimensional diffusion	$1 - (2x/3) - (1 - x)^{2/3}$
A2	Random nucleation, Avrami's equation	$[-\ln(1 - x)]^{1/2}$
A3	Random nucleation, Avrami's II equation	$[-\ln(1 - x)]^{1/3}$
A4	Random nucleation, Avrami's II equation	$[-\ln(1 - x)]^{1/4}$
R1	Linear controlled reaction, linear	x
R2	Phase interface, surface controlled reaction	$1 - (1 - x)^{1/2}$
R3	Phase boundary controlled reaction, volumetric	$1 - (1 - x)^{1/3}$
F1	Random nucleation, one embryo per molecule	$-\ln(1 - x)$
F2	Random nucleation, two embryos per molecule	$[1/(1 - x)] - 1$
F3	Random nucleation, three embryos per molecule	$[1/(1 - x)2] - 1$
Fn (n-order)	Random nucleation, n embryos per molecule	$[1/(1 - x)n - 1] - 1$

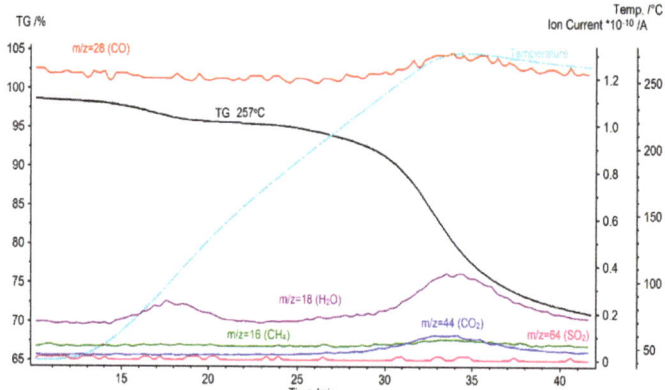

Figure A7. TG curve and ion current curves for the maize straw torrefaction process carried out up to 257.5 °C.

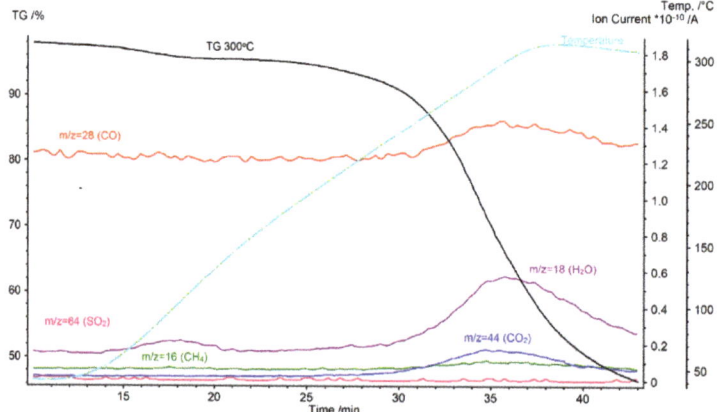

Figure A8. TG curve and ion current curves for the maize straw torrefaction process carried out up to 300 °C.

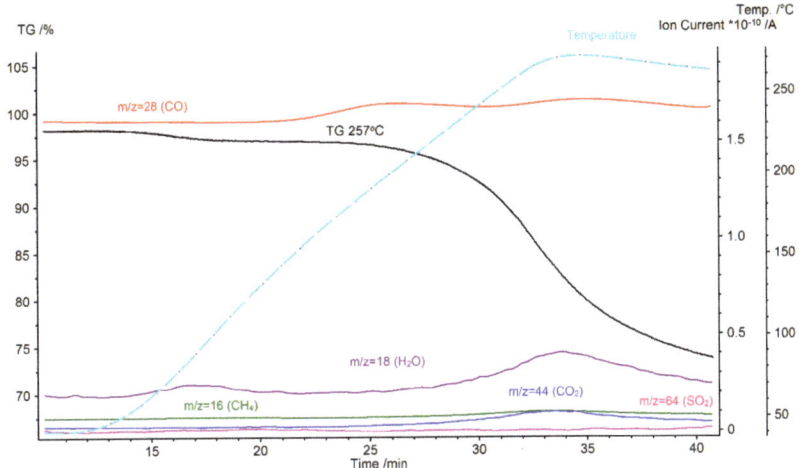

Figure A9. TG curve and ion current curves for the oat straw torrefaction process carried out up to 257.5 °C.

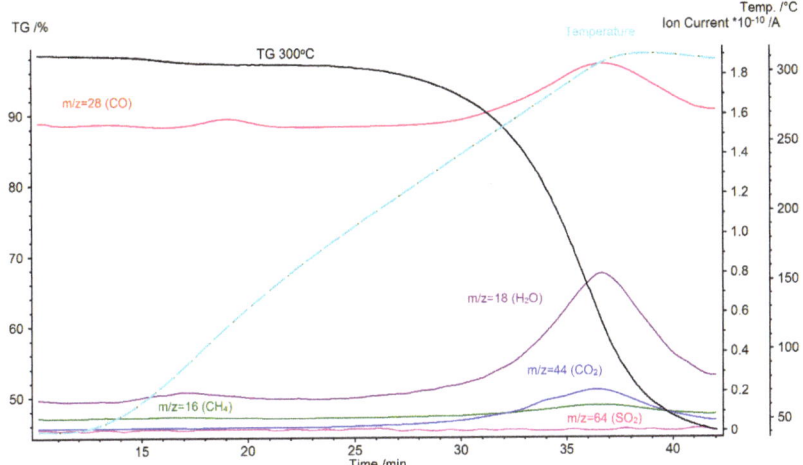

Figure A10. TG curve and ion current curves for the oat straw torrefaction process carried out up to 300 °C.

References

1. Adamczyk, F.; Frąckowiak, P.; Mielec, K.; Kośmicki, Z. Problematyka badawcza w procesie zagęszczania słomy przeznaczonej na opał. *J. Res. Appl. Agric. Eng.* **2005**, *50*, 5–8. (In Polish)
2. Poudel, J.; Oh, S.C. Effect of Torrefaction on the Properties of Corn Stalk to Enhance Solid Fuel Qualities. *Energies* **2014**, *7*, 5586–5600. [CrossRef]
3. Xing, X.; Fan, F.; Jiang, W. Characteristics of biochar pellets from corn straw under different pyrolysis temperatures. *R. Soc. Open Sci.* **2018**, *5*, 172346. [CrossRef] [PubMed]
4. Meia, Y.; Chea, Q.; Yangbc, Q.; Draperd, C.; Yanga, H.; Zhanga, S.; Chena, H. Torrefaction of different parts from a corn stalk and its effect on the characterization of products. *Ind. Crops Prod.* **2016**, *92*, 26–33. [CrossRef]
5. Gent, S.; Twedt, M.; Gerometta, C.; Almberg, E. *Theoretical and Applied Aspects of Biomass Torrefaction for Biofuels and Value-Added Products*, 1st ed.; Butterworth-Heinemann: Oxford, UK, 2017; ISBN 9780128094839.

6. Gładki, J. *Biowęgiel szansą dla zrównoważonego rozwoju [BIOCHAR as a Chance for Sustainable Development]*, 2nd ed.; Oficyna Poligrafi czna Apla Sp.J.: Sędziszów, Poland, 2017; pp. 60–85. ISBN 978-83-65487-05-06.
7. Gradziuk, P. Ekonomiczne i ekologiczne aspekty wykorzystania słomy na cele energetyczne. *Acta Agrophysica* **2006**, *8*, 18–44. (In Polish)
8. Ren, X.; Rokni, E.; Sun, R.; Meng, X.; Levendis, Y.A. Evolution of Chlorine-Bearing Gases during Corn Straw Torrefaction at Different Temperatures. *Energy Fuels* **2017**, *31*, 13713–13723. [CrossRef]
9. Gradziuk, P.; Kościk, K. Analiza możliwości i kosztów pozyskania biomasy na cele energetyczne na potrzeby energetycznego wykorzystania w gminie Clomas. *Opracowanie na zlecenie Urzędu Gminy Clomas* **2007**, 35–105. Available online: https://scholar.google.com/scholar?cluster=11289157253013309311&hl=en&oi=scholarr (accessed on 15 April 2020). (In Polish).
10. Tiffany, D.G.; Lee, W.F.; Morey, V.; Kaliyan, N. Economic Analysis of Biomass Torrefaction Plants Integrated with Corn Ethanol Plants and Coal-Fired Power Plants. *Adv. Energy Res.* **2013**, *1*, 127. [CrossRef]
11. Cruz, D.C. Production of Bio-coal and Activated Carbon from Biomass. Master's Thesis, The University of Western Ontario, London, Canada, 2012.
12. Denisiuk, W. Słoma—Potencjał masy i energii. *Inżynieria Rolnicza* **2008**, *2*, 23–30. (In Polish)
13. Grzybek, A.; Gradziuk, P.; Kowalczyk, K. *Słoma-Energetyczne Paliwo*; Wydaw: Warsaw, Poland, 2001. (In Polish)
14. Chen, D.; Cen, K.; Cao, X.; Li, Y.; Zhang, Y.; Ma, H. Restudy on torrefaction of corn stalk from the point of view of deoxygenation and decarbonization. *J. Anal. Appl. Pyrolysis* **2018**, *135*, 85–93. [CrossRef]
15. Meng, X.; Sun, R.; Zhou, W.; Liu, X.; Yan, Y.; Ren, X. Effects of corn ratio with pine on biomass co-combustion characteristics in a fixed bed. *Appl. Therm. Eng.* **2018**, *142*, 30–42. [CrossRef]
16. Ren, X.; Rokni, E.; Liu, Y.; Levendis, Y.A. Reduction of HCl Emissions from Combustion of Biomass by Alkali Carbonate Sorbents or by Thermal Pretreatment. *J. Energy Eng.* **2018**, *144*, 04018045. [CrossRef]
17. Shankar, J.; Timothy, T.; Christopher, K.; Wright, T.; Richard, D. Boardman, Proximate and Ultimate Compositional Changes in Corn Stover during Torrefaction Using Thermogravimetric Analyzer and Microwaves. In Proceedings of the ASABE Annual International Meeting, Dallas, TX, USA, 29 July–1 August 2012.
18. Almberg, E.R.; Twedt, M.P.; Gerometta, C.; Gent, S.P. Economic Feasibility of Corn Stover Torrefaction for Distributed Processing Systems. In Proceedings of the ASME 2016 10th International Conference on Energy Sustainability collocated with the ASME 2016 Power Conference and the ASME 2016 14th International Conference on Fuel Cell Science, Engineering and Technology, Charlotte, NC, USA, 26–30 June 2016.
19. Andini, A.; Bonnet, S.; Rousset, P.; Patumsawad, S.; Pattiya, A. Torrefaction study of Indonesian crop residues subject to open burning. In Proceedings of the Technology & Innovation for Global Energy Revolution, Bangkok, Thailand, 28–30 November 2018.
20. Lin, Y.-L. Effects of Microwave—Induced Torrefaction on Waste Straw Upgrading. *Int. J. Chem. Eng. Appl.* **2015**, *6*, 6. [CrossRef]
21. Ren, X.; Rokni, E.; Zhang, L.; Wang, Z.; Liu, Y.; Levendis, Y.A. Use of Alkali Carbonate Sorbents for Capturing Chlorine-Bearing Gases from Corn Straw Torrefaction. *Energy Fuels* **2018**, *32*, 11843–11851. [CrossRef]
22. Strzelczyk, M.; Steinhoff-Wrześniewska, A.; Helis, M. Biowęgiel dla rolnictwa alternatywne rozwiązanie w gospodarce odpadami. In Proceedings of the Konferencja naukowa Współczesne wyzwania gospodarki wodnej na obszarach wiejskich, Zdrój, Poland, 17–19 September 2018. (In Polish).
23. Saleh, S.B. *Torrefaction of Biomass for Power Production*; Technical University of Denmark: Lyngby, Denmark, 2013.
24. ISO 16994. *Solid Biofuels—Determination of Total Content of Sulfur and Chlorine*; International Organization for Standardization: Geneva, Switzerland, 2016.
25. ISO 16948. *Solid Biofuels—Determination of Total Content of Carbon, Hydrogen and Nitrogen*; International Organization for Standardization: Geneva, Switzerland, 2015.
26. ISO 18122. *Solid Biofuels—Determination of Ash Content*; International Organization for Standardization: Geneva, Switzerland, 2015.
27. ISO 18123. *Solid Biofuels—Determination of the Content of Volatile Matter*; International Organization for Standardization: Geneva, Switzerland, 2015.
28. ISO 18125. *Solid Biofuels—Determination of Calorific Value*; International Organization for Standardization: Geneva, Switzerland, 2017.

29. Bates, R.B.; Ghoniem, A.F. Biomass torrefaction: Modeling of volatile and solid product evolution kinetics. *Bioresour. Technol.* **2012**, *124*, 460–469. [CrossRef] [PubMed]
30. Kruczek, H.; Wnukowski, M.; Niedzwiecki, Ł.; Guziałowska-Tic, J. Torrefaction as a Valorization Method Used Prior to the Gasification of Sewage Sludge. *Energies* **2019**, *12*, 175. [CrossRef]
31. Standard Test Method for Kinetic Parameters for Thermally Unstable Materials Using Differential Scanning Calorimetry and the Flynn/Wall/Ozawa Method Active Standard ASTM E698 | Developed by Subcommittee: E37.01, Book of Standards Volume: 14.05. Available online: https://td.chem.msu.ru/uploads/files/courses/special/specprac-ta/lit/ASTM%20E%20698%20-%2016.pdf (accessed on 15 April 2020).
32. Syguła, E.; Koziel, J.A.; Białowiec, A. Proof-of-Concept of Spent Mushrooms Compost Torrefaction—Studying the Process Kinetics and the Influence of Temperature and Duration on the Calorific Value of the Produced Biocoal. *Energies* **2019**, *12*, 3060. [CrossRef]
33. Szufa, S.; Dzikuć, M.; Adrian, Ł.; Piersa, P.; Romanowska-Duda, Z.; Marczak, M.; Błaszczuk, A.; Piwowar, A.; Lewandowska, W. Torrefaction of oat straw to use as solid biofuel, an additive to organic fertilizers for agriculture purposes and activated carbon—TGA analysis, kinetics. *E3S Web Conf.* **2020**, *154*, 02004. [CrossRef]
34. Romanowska-Duda, Z.; Piotrowski, K.; Wolska, B.; Dębowski, M.; Zieliński, M.; Dziugan, P.; Szufa, S. Stimulating effect of ash from Sorghum on the growth of Lemnaceae—A new source of energy biomass. In *Renewable Energy Sources: Engineering, Technology, Innovation, Springer Proceedings in Energy*; Wróbel, M., Jewiarz, M., Szlęk, A., Eds.; Springer Nature: Cham, Switzerland, 2019; pp. 341–349. ISBN 978-3-030-13887-5. [CrossRef]
35. Szufa, S.; Adrian, Ł.; Piersa, P.; Romanowska-Duda, Z.; Ratajczyk-Szufa, J. Torrefaction process of millet and cane using batch reactor. In *Renewable Energy Sources: Engineering, Technology, Innovation, Springer Proceedings in Energy*; Wróbel, M., Jewiarz, M., Szlęk, A., Eds.; Springer Nature: Cham, Switzerland, 2019; pp. 371–379. ISBN 978-3-030-13887-5. [CrossRef]
36. Adrian, Ł.; Szufa, S.; Piersa, P.; Kurowski, K. Experimental research and simulation of computer processes of heat exchange in a heat exchanger working on the basis of the principle of heat pipes for the purpose of heat transfer from the ground. In Proceedings of the 4th Renewable Energy Sources—Research and Business RESRB 2019 Conference, Wrocław, Poland, 8–9 July 2019.
37. Adrian, Ł.; Szufa, S.; Piersa, P.; Romanowska-Duda, Z.; Grzesik, M.; Cebula, A.; Kowalczyk, S.; Ratajczyk-Szufa, J. Thermographic analysis and experimental work using laboratory installation of heat transfer processes in a heat pipe heat exchanger utilizing as a working fluid R404A and R407A. In *Renewable Energy Sources: Engineering, Technology, Innovation, Springer Proceedings in Energy*; Wróbel, M., Jewiarz, M., Szlęk, A., Eds.; Springer Nature: Cham, Switzerland, 2019; pp. 799–807. ISBN 978-3-030-13887-5. [CrossRef]
38. Szufa, S.; Adrian, Ł.; Piersa, P.; Romanowska-Duda, Z.; Grzesik, M.; Cebula, A.; Kowalczyk, S. Experimental studies on energy crops torrefaction process using batch reactor to estimate torrefaction temperature and residence time. In *Renewable Energy Sources: Engineering, Technology, Innovation*; Springer: Cham, Switzerland, 2018; pp. 365–373. ISBN 978-3-319-72370-9. [CrossRef]
39. Adrian, Ł.; Szufa, S.; Piersa, P.; Romanowska-Duda, Z.; Grzesik, M.; Cebula, A.; Kowalczyk, S. Experimental research and thermographic analysis of heat transfer processes in a heat pipe heat exchanger utilizing as a working fluid R134A. In *Renewable Energy Sources: Engineering, Technology, Innovation*; Springer: Cham, Switzerland, 2018; pp. 413–421. ISBN 978-3-319-72370-9. [CrossRef]
40. Szufa, S.; Romanowska-Duda, B.Z.; Grzesik, M. Torrefaction proces of the Phragmites Communis growing in soil contaminated with cadmium. In Proceedings of the 20th European Biomass Conference and Exibition, Milan, Italy, 18–22 June 2014; pp. 628–634, ISBN 978-88-89407-54-7.
41. Dzikuć, M.; Kułyk, P.; Dzikuć, M.; Urban, S.; Piwowar, A. Outline of Ecological and Economic Problems Associated with the Low Emission Reductions in the Lubuskie Voivodeship (Poland). *Polish J. Environ. Stud.* **2019**, *28*, 65–72. [CrossRef]
42. Adamczyk, J.; Piwowar, A.; Dzikuć, M. Air protection programmes in Poland in the context of the low emission. *Environ. Sci. Pollut. Res.* **2017**, *24*, 16316–16327. [CrossRef]
43. Dzikuć, M.; Adamczyk, J.; Piwowar, A. Problems associated with the emissions limitations from road transport in the Lubuskie Province (Poland). *Atmos. Environ.* **2017**, *160*, 1–8. [CrossRef]
44. Dzikuć, M. Problems associated with the low emission limitation in Zielona Góra (Poland): Prospects and challenges. *J. Clean. Prod.* **2017**, *166*, 81–87. [CrossRef]

45. Dzikuć, M.; Tomaszewski, M. The effects of ecological investments in the power industry and their financial structure: A case study for Poland. *J. Clean. Prod.* **2016**, *118*, 48–53. [CrossRef]
46. Piwowar, A.; Dzikuć, M. LCA w produkcji agrochemikaliów. Procedura, kategorie wpływu, możliwości wykorzystania. *Przemysł Chemiczny* **2017**, *96*, 271–274. (In Polish) [CrossRef]
47. Marczak, M.; Karczewski, M.; Makowska, D.; Burmistrz, P. Impact of the temperature of waste biomass pyrolysis on the quality of the obtained biochar. *Agric. Eng.* **2016**, *20*, 115–124.
48. Junga, R.; Pospolita, J.; Niemiec, P. Combustion and grindability characteristics of palm kernel shells torrefied in a pilot-scale installation. *Renew. Energy* **2020**, *147*, 1239–1250. [CrossRef]

© 2020 by the authors. Licensee MDPI, Basel, Switzerland. This article is an open access article distributed under the terms and conditions of the Creative Commons Attribution (CC BY) license (http://creativecommons.org/licenses/by/4.0/).

Article

Selective Hydrogenation of Phenol to Cyclohexanol over Ni/CNT in the Absence of External Hydrogen

Changzhou Chen [1,2], Peng Liu [1,2], Minghao Zhou [1,2,3,*], Brajendra K. Sharma [3,*] and Jianchun Jiang [1,2,*]

1. Institute of Chemical Industry of Forest Products, Chinese Academy of Forestry, Key Laboratory of Biomass Energy and Material, Jiangsu Province, National Engineering Laboratory for Biomass Chemical Utilization, Key and Open Laboratory on Forest Chemical Engineering, SFA, Nanjing 210042, China; changzhou_chen@163.com (C.C.); liupengnl@163.com (P.L.)
2. Co-Innovation Center of Efficient Processing and Utilization of Forest Resources, Nanjing Forestry University, Nanjing 210037, China
3. Illinois Sustainable Technology Center, Prairie Research Institute, one Hazelwood Dr., Champaign, University of Illinois at Urbana-Champaign, Champaign, IL 61820, USA
* Correspondence: zmhzyk19871120@163.com (M.Z.); bksharma@illinois.edu (B.K.S.); jiangjc@icifp.cn (J.J.)

Received: 7 January 2020; Accepted: 11 February 2020; Published: 14 February 2020

Abstract: Transfer hydrogenation is a novel and efficient method to realize the hydrogenation in different chemical reactions and exploring a simple heterogeneous catalyst with high activity is crucial. Ni/CNT was synthesized through a traditional impregnation method, and the detailed physicochemical properties were performed by means of XRD, TEM, XPS, BET, and ICP analysis. Through the screening of loading amounts, solvents, reaction temperature, and reaction time, 20% Ni/CNT achieves an almost complete conversion of phenol after 60 min at 220 °C in the absence of external hydrogen. Furthermore, the catalytic system is carried out on a variety of phenol derivatives for the generation of corresponding cyclohexanols with good to excellent results. The mechanism suggests that the hydrogenation of phenol to cyclohexanone is the first step, while the hydrogenation of cyclohexanone for the generation of cyclohexanol takes place in a successive step. Moreover, Ni/CNT catalyst can be magnetically recovered and reused in the next test for succeeding four times.

Keywords: phenol; hydrogenation; Ni/CNT; cyclohexanol; transfer hydrogenation

1. Introduction

Cyclohexanol is an important chemical raw materials, for example, it could be used as the main intermediate for the production of adipic acid, hexamethylene diamine, cyclohexanone, and caprolactam, or as an excellent solvent for rubber, resin, nitro fiber, and metal soap [1–5]. Except for this, its downstream product, cyclohexane, can also be converted into various useful chemicals [6–8]. Hence, the production of cyclohexanol has attracted increasing attention, both in industry and in the laboratory, recently. In the past several years, tremendous efforts have been devoted to the hydrogenation of phenol for the generation of cyclohexanol. The main pathways can be classified into two groups, gaseous [9] and liquid phase [10]. In the gaseous phase, a variety of supported noble metals, such as Pd [11,12], Pt [13,14], Ru [15,16], Rh [17,18], have been reported to be effective in the hydrogenation system. For example, Wang et al. reported a hierarchically porous ZSM-5 zeolite with micropore and b-axis-aligned mesopore-supported Ru nanoparticles (Ru/HZSM-5-OM), which were highly effective for the hydrogenation of both phenol and its derivatives to the corresponding cyclohexane [15]. However, elevated temperature and high pressure of hydrogen (150 °C, 4 MPa) were required, and carbonaceous deposits in the process would result in the deactivation of catalyst. In the liquid phase, although the hydrogenation process could be performed under a relatively lower temperature [19], the main shortcoming of recent hydrogenation systems is that they could not avoid the presence of hydrogen [20]. For instance, Li et al. developed a series of Pd@FDU-N catalysts for the hydrogenation of phenol in the presence of 0.1 MPa H_2 [21]. Therefore, the exploration of a cheap and effective catalyst for the hydrogenation of phenol in the absence of hydrogen is essential. Catalytic transfer hydrogenation (CTH) has been regarded as a good alternative to avoid the application of high-pressure hydrogen [22–25]. Galkin et al. reported that Pd/C could catalyze transfer hydrogenolysis of β-O-4 model compound in lignin employing formic acid as a hydrogen-donor for the generation of acetophenone and phenol derivatives [22]. Paone et al. reported the transfer hydrogenolysis of α-O-4 model compound in lignin conducted on Pd/Fe_3O_4 under 240 °C [23]. Wu et al. found that Ru/C could efficiently catalyze the cleavage of the 4-O-5 aromatic ether bond in a variety of lignin-derived compounds through the transfer hydrogenolytic pathway using isopropanol as the hydrogen-donor solvent [24]. Despite those achievements for the transfer hydrogenolytic transformation of biomass to value-added chemicals, the catalytic hydrogenation without the use of external hydrogen remains a great challenge [25,26].

Carbon nanotubes (CNT) are interesting materials for carrying out chemical reactions in confined spaces, and several organic reactions have been carried out [27–29] inside the nanotubes. In this work, CNT supported nickel catalyst was introduced into the hydrogenation of phenol for the generation of cyclohexanol via a transfer hydrogenolytic route. Through the screening of loading amounts, solvents, temperature and time, herein we reported a mild reaction condition (20% Ni/CNT, iPrOH, 220 °C and 60 min) for the hydrogenation of phenol to cyclohexanol. A variety of phenol derivatives were also conducted to verify the high efficiency of this catalytic reaction system. The possible reaction mechanism was finally investigated as well. The detailed physicochemical properties were investigated by means of XRD, TEM, XPS, BET, and ICP analysis.

2. Experimental

Materials: CNT was purchased from Aladdin Industrial Inc. Shanghai, China and pre-treated in HNO_3 before use [27,28]. Ni(NO_3)$_2$·$6H_2O$ was provided by Aladdin Industrial Inc. Shanghai, China. Phenol and other derivatives were obtained from tansoole.com. All chemicals were obtained from commercial sources and used without further purification.

General procedure for Ni/CNT catalyzed phenol: In one typical catalytic reaction process, 500 mg of phenol and other derivatives, 50 mg of Ni/CNT catalyst, and 10 mL isopropanol were placed in a 25 mL stainless steel reactor. After being sealed, the catalytic reaction was stirred at desired temperature for desired time. The reactor was naturally cooled to room temperature after the reaction. The mixture solution was filtered to collect the catalyst and the filtrate was analyzed by the Gas Chromatograph/Mass Spectrometer (GC/MS, Agilent 7890) utilizing n-dodecane as an internal standard. The collected catalyst was washed with isopropanol three times and dried at 105 °C for the next test under the optimal reaction conditions. The conversion and product yields in the liquid phase were calculated according to the following formula, respectively:

$$Conversion = \frac{mole\ of\ reacted\ substrate}{total\ mole\ of\ substrate\ feed} \times 100\%$$

$$Yield\ of\ cyclohexanone = \frac{mole\ of\ cyclohexanone}{total\ mole\ of\ substrate\ feed} \times 100\%$$

$$Yield\ of\ cyclohexanol = \frac{mole\ of\ cyclohexanol}{total\ mole\ of\ substrate\ feed} \times 100\%$$

Catalyst preparation: Ni/CNT was prepared by an impregnation method. In one typical process of 20% Ni/CNT catalyst, Ni(NO_3)$_2$·$6H_2O$ (0.2 g) was dissolved in 30 mL of deionized water. Then, CNT (2.0 g) was added to the above solution and stirred for 24 h until a uniform dispersion. The obtained suspension was dried for 12 h at 105 °C in the oven. Subsequently, the obtained black solid was calcined at 500 °C for 2 h in the muffle furnace, and finally reduced in a tube furnace under hydrogen atmosphere at 500 °C for 2 h. The preparation of Fe/CNT, Co/CNT, Mo/CNT was the same as Ni/CNT.

Catalyst characterization: Powder X-ray diffraction (XRD) was examined on a Bruker D8 Advance X-ray powder diffractometer. Transmission electron microscopy (TEM) images were tested using a TEM Tecnai G2 20. The X-ray photoelectron spectroscopy (XPS) was carried out on an ESCALAB-250 (Thermo-VG Scientific, USA) spectrometer with Al Kα (1486.6 eV) irradiation source. The textural properties of the Ni/CNT catalyst were tested by N_2 adsorption–desorption isotherms using a COULTER SA 3100 analyzer, and the Brunauer–Emmett–Teller (BET) surface area was evaluated by the N_2 adsorption–desorption isotherms.

2.1. Catalyst Characterization

Figure 1 shows the XRD patterns of two different samples, including solo CNT, and the optimal catalyst in the system for the hydrogenation of phenol. The peak at 26.5° could be assigned to the diffraction peaks of the (002) planes of the graphite-like tube-wall of the CNT [30–32]. Apart from the CNT feature, the presence of metallic Ni in the reduced 20% Ni/CNT catalyst was clearly revealed. As could be clearly seen from the XRD results, only one kind of Ni-based phase was observed for Ni/CNT catalyst. Peaks at 44.5°, 51.9°, and 76.4° could be assigned to the diffraction of the (111), (200), and (220) planes of metallic Ni [27], indicating that only metallic nickel species were observed in Ni/CNT catalyst.

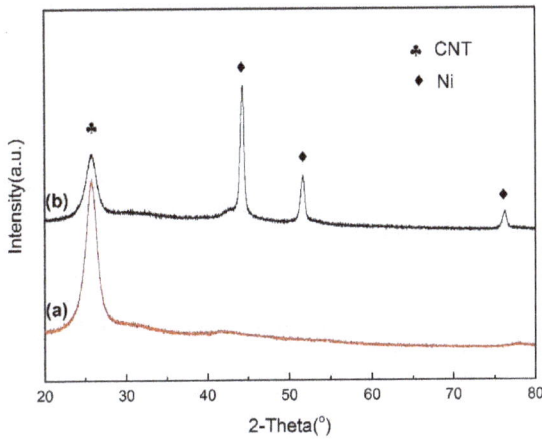

Figure 1. Characterization of the 20% Ni/CNT catalyst. (a) XRD pattern of CNT, (b) XRD pattern of 20% Ni/CNT.

XPS detection was carried out to analyze the composition of prepared 20% Ni/CNT. The absorption peaks were identified for C, O, and Ni. As for the presence of oxygen element in Figure 2b, perhaps it was due to the presence of oxygen-containing groups of CNT during the HNO_3 pre-processing [30,31]. In addition, the oxidation of Ni/CNT catalyst before or during the XPS examination could also lead to the presence of oxygen element. The binding energies at 852.9 and 856.3 eV were observed for 20% Ni/CNT (Figure 2a), which corresponded to Ni^0 (2p 3/2) and Ni^{2+} (2p3/2), respectively. The binding energies at 872.3 and 874.8 eV corresponded to the main lines of Ni^0 (2p 1/2) and Ni^{2+} (2p 1/2). It could be found that part of metallic Ni was oxidized before or during the XPS testing, which contributed to the understanding of Ni^{2+} detection in the XPS spectra.

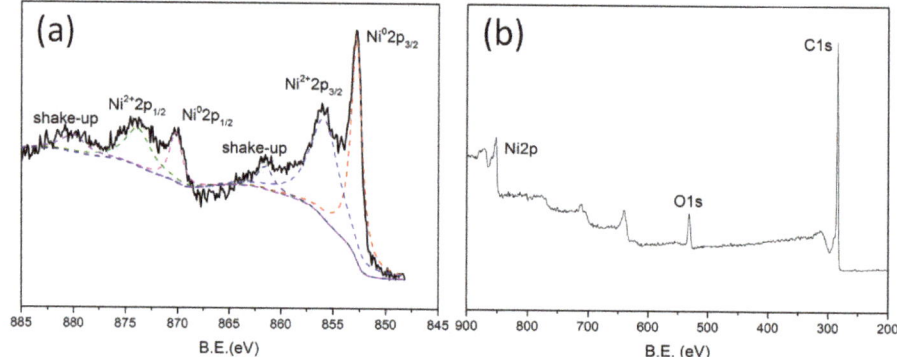

Figure 2. Characterization of the 20% Ni/CNT catalyst. (**a**) XPS of Ni 2p, (**b**) XPS pattern of 20% Ni/CNT.

As can be seen in Figure 3, most of Ni particles displayed outside the CNT uniformly. However, the Ni particle size and its distribution varied from each other mainly owing to the loading amount of Ni on the surface of CNT. We selected 20% Ni/CNT catalyst as an example. The histogram of 20% Ni/CNT in given in Figure 3a, in which the mean size was 14.0 nm. Taken together, nanoparticle Ni on the surface of CNT performed an excellent activity in the hydrogenation of phenol under our reaction system.

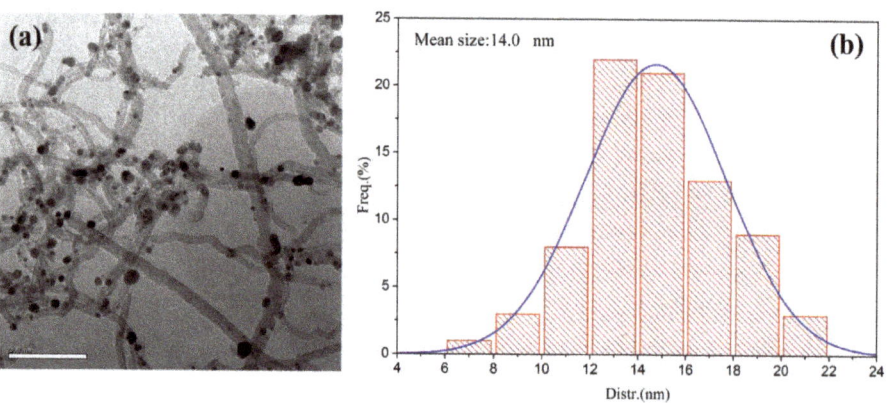

Figure 3. Characterization of the 20% Ni/CNT catalyst. (**a**) TEM image of 20% Ni/CNT, (**b**) Distribution of Ni particles.

Table 1 show the chemical and physical properties for the different ratios of Ni/CNT catalysts, including the Ni loading amounts, BET surface area, and average Ni-particle size. The ICP results prove that the metal loading amounts were in great accordance with the theoretical calculating value with the preparation of the catalysts. The BET surface area of 5% Ni/CNT catalyst was much larger than 25% Ni/CNT (range from 149.3 m^2/g to 120.5 m^2/g), suggesting that Ni particles could be better dispersed on the surface of CNT. Owing to the different loading amounts of Ni on the CNT, the average particle size of Ni/CNT with different metal amounts ranged from 12.1 to 15.6 nm (estimated from XRD in Table 1). It proved that excessive metal loading on CNT could inevitably lead to metal agglomeration, resulting in an increase of the metal particle size [32]. The TEM also gave the same results on the distribution of Ni particles (Table 1). Hence, these changes in catalyst properties would directly influence the catalytic activity in the hydrogenation of phenol, which will be discussed in detail in the following experiments.

Table 1. Chemical and physical properties of CNT-supported Ni catalysts.

Catalysts	Composition (wt.%) [a] Ni	S_{BET} [b] (m^2/g)	Average Metal Size [c] (nm)	Average Metal Size [d] (nm)
CNT	/	180.6	/	/
5% Ni/CNT	4.32	149.3	12.1	12.8
10% Ni/CNT	10.55	145.6	12.6	13.1
15% Ni/CNT	14.66	135.8	12.8	13.5
20% Ni/CNT	19.68	132.4	13.1	14.0
25% Ni/CNT	24.09	120.5	15.6	16.5

[a]—measured by ICP analysis, [b]—evaluated from N_2 adsorption-desorption isotherms, [c]—estimated by XRD, [d]—measured by TEM.

2.2. Activity of Various Catalysts for the Hydrogenation of Phenol

First of all, hydrogenation of phenol was chosen as a typical model reaction to explore the catalytic performance over a variety of catalysts (Table 2). When carried out without any catalysts in the hydrogenolytic system, the reaction failed to transform phenol to cyclohexanol (Table 2, entry 1). Meanwhile, solo CNT also showed a poor activity under our reaction condition (Table 2, entry 2), indicating that CNT only served as a support, and showed no catalytic effect in the catalytic process. Then, we tried different catalysts for the transformation of phenol, including 10% Fe/CNT, 10% Co/CNT, 10% Mo/CNT and 10% Ni/CNT (Table 2, entries 3–5 and entry 7). It was surprising to discover that the supported Ni catalyst exhibited good performance in the hydrogenation of phenol to generate cyclohexanol, indicating that the substrate phenol adsorbed on the metal surface prior to the reaction decreased the activation barrier of hydrogenation reaction [33]. The catalytic activity could be sorted as follows: 10% Ni/CNT > 10% Co/CNT > 10% Mo/CNT > 10% Fe/CNT. However, the loading amount of Ni on the surface of CNT could also affect the efficiency of the hydrogenolytic process. It could be clearly seen in Table 2 that the conversion increased with the increased loading amount of Ni (from 5% to 25%), and the trend of the cyclohexanol yield showed the same result. Meanwhile, we found that the yield of cyclohexanol reached 95% when 20% Ni/CNT was used in our hydrogenolytic system (Table 2, entries 6–10). On the base of the results above, 20% Ni/CNT was chosen as the most suitable catalyst for the subsequent exploration.

Table 2. Optimization of catalysts for the hydrogenation of phenol [a].

Entry	Catalyst	T.(°C)/t.(h)	Con. (%)	Yield (%) [b]	
				2a	3a
1	none	180/4	0	0	0
2	CNT	180/4	0	0	
3	10% Fe/CNT	180/4	8	2	0
4	10% Co/CNT	180/4	48	5	5
5	10% Mo/CNT	180/4	27	4	22
6	5% Ni/CNT	180/4	73	3	66
7	10% Ni/CNT	180/4	81	4	68
8	15% Ni/CNT	180/4	95	3	88
9	20% Ni/CNT	180/4	100	2	95
10	25% Ni/CNT	180/4	100	2	94

[a] Reaction conditions: **1a** (500 mg), catalyst (50 mg), Isopropanol (10 mL). [b] The conversion and yield were determined by GC/MS with n-dodecane as the internal standard.

2.3. Influence of Hydrogen-Donor Solvents

Alcohol, as an excellent hydrogen-donor solvent, plays an important role in the catalytic transfer hydrogenolytic process. Therefore, we selected three most common alcohols in our catalytic system. Unfortunately, methanol and ethanol failed the transformation of phenol to cyclohexanol (Table 3, entries 2–3). H_2O was considered as a green and environmentally friendly solvent, and always applied in various chemical reactions. However, 20% Ni/CNT catalyst showed a poor activity in the hydrogenation of phenol in aqueous media (Table 3, entry 4). When isopropanol was utilized in our catalytic system, a satisfactory result was achieved that cyclohexanol was obtained in the yield of 95% (Table 3, entry 1). Therefore, isopropanol was the most efficient hydrogen-donor solvent for the transformation of phenol to cyclohexanol.

Table 3. Optimization of solvents for the hydrogenation of phenol [a].

Entry	Solvent	T.(°C)/ t.(h)	Con. (%)	Yield (%) [b]	
				2a	3a
1	isopropanol	180/4	100	2	95
2	methanol	180/4	15	3	6
3	ethanol	180/4	13	2	5
4	H_2O	180/4	5	0	0

[a] Reaction conditions: 1a (500 mg), 20% Ni/CNT (50 mg), Solvent (10 mL). [b] The conversion and yield were determined by GC/MS with n-dodecane as the internal standard.

2.4. Influence of Reaction Temperature and Reaction Time

In addition, reaction temperature and reaction time were important key factors for the transformation of phenol (Figure 4). As shown in Figure 4a, the conversion of phenol progressively increased when the reaction temperature increased from 160 to 220 °C, and phenol could almost totally transfer to cyclohexanol at 220 °C with a reaction time of 60 min. In the catalytic process, the dominant products were cyclohexanol, with a small amount of cyclohexanone detected in the GC/MS. As the temperature continued to rise, the yield of cyclohexanol kept around 95%. Subsequently, it could be found that the conversion of phenol increased with the prolonging of the reaction time (Figure 4b), and the phenol was completely consumed in only 60 min at 220 °C. With the prolonged reaction time, the yield of cyclohexanol increased when the reaction time increased from 20 to 60 min. When the reaction time exceeded 60 min, the conversion of phenol achieved 100%, and the yield of cyclohexanol kept at the highest point. Through the screening of temperature and time of the hydrogenation of phenol, the optimal temperature and time could be summarized as 220 °C and 60 min, respectively.

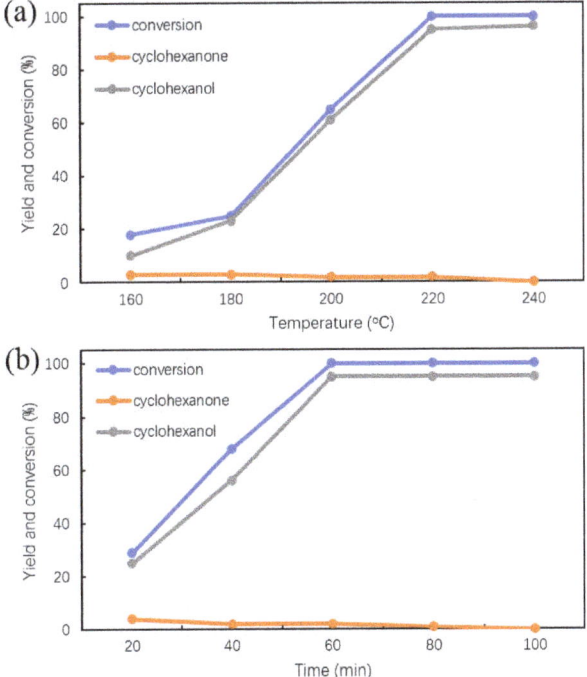

Figure 4. Influence of reaction temperature (**a**) and time (**b**) on the hydrogenation of phenol. Reaction conditions: Phenol (500 mg), 20% Ni/CNT (50 mg), isopropanol (10 mL); reaction time, 60 min (for a); reaction temperature, 220 °C (for b).

2.5. Scope of the Substrates

Encouraged by the perfect performance of the efficient hydrogenation of phenol, a variety of phenol derivatives were tested on 20% Ni/CNT. As shown in Table 4, phenol could be completely transformed to afford the corresponding cyclohexanol in a yield of 95% without additional hydrogen under the optimal reaction condition (Table 4, entry 1). Methylphenol derivatives gave methylcyclohexanols with a poor conversion and yield under 220 °C at a reaction time of 60 min. Therefore, we increased the temperature from 220 °C to 240 °C. As excepted, methylphenol derivatives gave methylcyclohexanols with both perfect conversion and yield (Table 4, entries 2–4). Hydroxy substituted phenol could not achieve the CTH process under optimal conditions as well. Hence, 240 °C was employed in our reaction system and a 79% yield of cyclohexane-1,2-diol was obtained in the reaction time of 60 min (Table 4, entries 5). Based on the relevant literature [34], the reason for the small decrease in the yields of corresponding cyclohexanol derivatives probably is the steric effect. Hence, the presence of substituent imposes a steric repulsion, which might lower the degree of adsorption and thus hydrogenation rate. Taken together, the transfer hydrogenation system was confirmed suitable for the hydrogenation of diverse phenol derivatives with good to excellent results.

Table 4. Hydrogenation of phenol with substituted functional groups with 20% Ni/CNT in isopropanol [a].

Entry	Substrate	T (°C)/t (min)	Conversion (%)	Yield of Product (%)
1	phenol	220/60	100	cyclohexanol 95
2	2-methylphenol	240/60	89	2-methylcyclohexanol 80
3	3-methylphenol	240/60	90	3-methylcyclohexanol 82
4	4-methylphenol	240/60	85	4-methylcyclohexanol 78
5	2,6-dimethylphenol	240/60	88	2,6-dimethylcyclohexanol 79

[a] Reaction condition: 20% Ni/CNT (50 mg), substrate (500 mg), solvent (10 mL).

2.6. Mechanism Studies and the Recyclability of Ni/CNT

The mechanism of the hydrogenation of phenol over 20% Ni/CNT was proposed in Figure 6. According to the time courses of phenol and products (Figure 4), the hydrogenation of phenol proceeded via cyclohexanone. In order to corroborate this, we investigated the activity of cyclohexanol and cyclohexanone in the process of transfer hydrogenation. When solo cyclohexanol was carried out in our catalytic system, no cyclohexanol was transformed into cyclohexanone (Figure 5a). The results above show the hydrogenation of phenol to cyclohexanol via cyclohexanone under a transfer hydrogenation process. It can be seen from Figure 5b that, from the beginning of the process, the amount of cyclohexanone decreased and cyclohexanol was observed. Cyclohexanone was completely consumed after 30 min and cyclohexanol was achieved in high yield (over 90%). Apart from this, acetone was also detected in the catalytic process. In Table 5, we could find that 95% of cyclohexanol, 2% of cyclohexanone, 5.8 equivalents of acetone and 33.6 equivalents of isopropanol were observed for each equivalent of phenol consumed, which suggested that the hydrogenation of phenol was a stoichiometric reaction and each phenol molecule needed six isopropanol molecules for the generation of cyclohexanol. Therefore, the detailed mechanism could be summarized in Figure 6. The molecule adsorption of isopropanol occurred on the catalyst surface to produce H* and followed by phenol adsorption (i). Then, irreversible phenol hydrogenation took place for the generation of cyclohexanone, in which four equivalents of H* were consumed (ii–iv). In the final step, two equivalents of H* was employed in the catalytic process to transfer cyclohexanone to cyclohexanol (v). Based on the study above, we propose the mechanism in two steps: (1) Conversion of phenol to cyclohexanone over the Ni/CNT catalyst. (2) Transformation of cyclohexanol from cyclohexanone via keto-enol tautomerism.

Table 5. Distribution of products with 20% Ni/CNT in isopropanol.

Catalyst	Temperature/Time	Conversion	cyclohexanol	cyclohexanone	acetone	isopropanol
Ni/CNT	220 °C/60 min	100%	95%	2%	5.8 equiv	33.6 equiv

Reaction condition: 500 mg phenol (1.0 equiv), 50 mg 20% Ni/CNT, 10mL isopropanol (40.0 equiv). Yield was determined by GC/MS with n-dodecane as the internal standard.

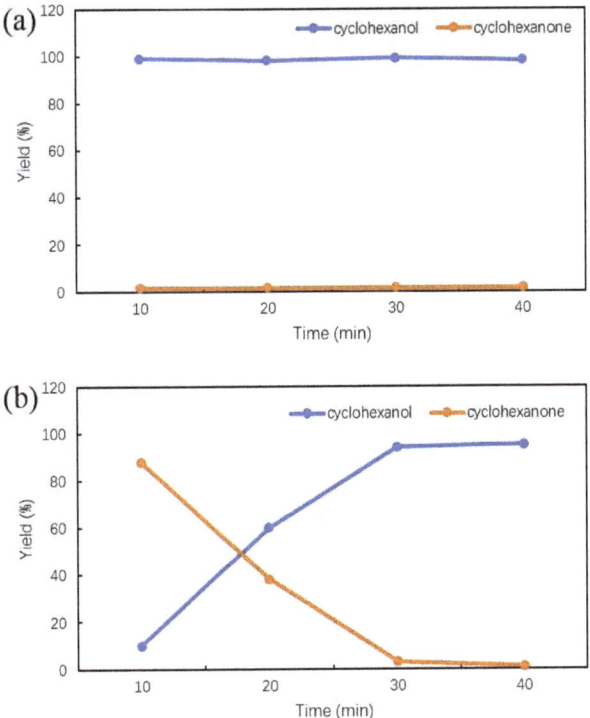

Figure 5. Time course of the yield of cyclohexanol carried out in the transfer hydrogenation (**a**). Time course of the yield of cyclohexanone carried out in the transfer hydrogenation (**b**).

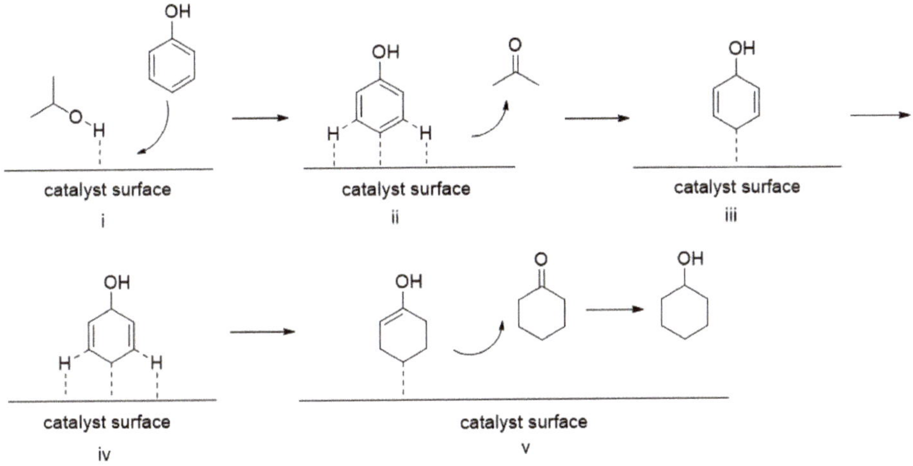

Figure 6. Catalytic pathway in the transfer hydrogenation of phenol.

In addition, the Ni/CNT catalyst could be magnetically recovered after a simple process (washing with isopropanol and dried over 105 °C) for the subsequent recycling tests. The recycling tests for the Ni/CNT catalyst in the transfer hydrogenation of phenol under the optimal reaction condition was subsequently evaluated. Ni/CNT maintained its activity for four succeeding runs, and no obvious decrease in cyclohexanol yield was observed, suggesting the good ability of the catalyst (Figure 7b).

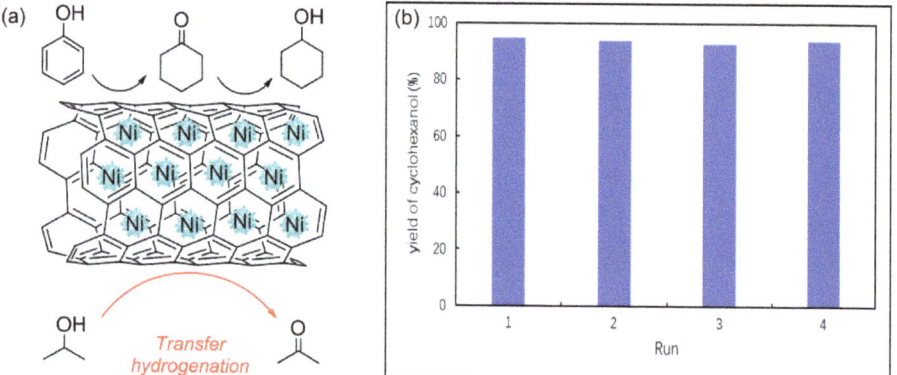

Figure 7. (**a**) Possible catalytic pathway in the transfer hydrogenation of phenol. (**b**) Result of recycling tests for Ni/CNT catalyst in the transfer hydrogenation of phenol.

3. Conclusions

In this paper, phenol was effectively hydrogenated to cyclohexanol by 20% Ni/CNT catalyst under a mild condition (220 °C, 60 min) using isopropanol as the hydrogen-donor solvent. The Ni/CNT catalyst not only had an excellent ability to catalyze the transfer hydrogenolytic cleavage of phenol and diverse phenol derivatives but also could be easily recovered magnetically from the reaction process for the next four recycling tests. Four succeeding recycling tests in the transfer hydrogenation of phenol proved the good stability of the Ni/CNT catalyst. The possible mechanism for the hydrogenation of phenol suggested that the possible pathway might proceed via the formation of cyclohexanone. This work will inspire various studies on the hydrogenation process in the absence of additional hydrogen.

Author Contributions: Conceptualization, M.Z. and J.J.; methodology, M.Z.; software, J.J.; validation, M.Z., B.K.S. and J.J.; formal analysis, C.C.; investigation, C.C.; resources, P.L.; data curation, P.L.; writing—original draft preparation, C.C.; writing—review and editing, M.Z.; visualization, M.Z.; supervision, J.J.; project administration, M.Z.; funding acquisition, M.Z. All authors have read and agreed to the published version of the manuscript.

Funding: This research was funded by the Fundamental Research Funds of CAF, grant number CAFYBB2018QB007 and the National Natural Science Foundation of China, grant number 31700645.

Acknowledgments: Authors are grateful for the financial support from the Fundamental Research Funds of CAF (No. CAFYBB2018QB007) and the National Natural Science Foundation of China (31700645).

Conflicts of Interest: The authors declare no conflict of interest.

Abbreviations

CNT	Carbon nanotube
ICP	Inductive Coupled Plasma Emission Spectrometer
XRD	Powder X-ray diffraction
XPS	X-Ray photoelectron spectroscopy
TEM	Transmission electron microscopy
BET	Brunauer- Emmett- Teller
GC/MS	Gas Chromatograph/Mass Spectrometer

References

1. Xiang, Y.; Ma, L.; Lu, C.; Zhang, Q.; Li, X. Aqueous system for the improved hydrogenation of phenol and its derivatives. *Green Chem.* **2008**, *10*, 939–943. [CrossRef]
2. Xiang, Y.; Kong, L.; Xie, P.; Xu, T.; Wang, J.; Li, X. Carbon nanotubes and activated carbons supported catalysts for phenol in situ hydrogenation: Hydrophobic/hydrophilic effect. *Ind. Eng. Chem. Res.* **2014**, *53*, 2197–2203. [CrossRef]
3. Porwal, G.; Gupta, S.; Sreedhala, S.; Elizabeth, J.; Khan, T.S.; Haider, M.A.; Vinod, C.P. Mechanistic insights into the pathways of phenol hydrogenation on Pd nanostructures. *ACS Sustain. Chem. Eng.* **2019**, *7*, 17126–17136. [CrossRef]
4. Song, Z.; Ren, D.; Wang, T.; Jin, F.; Jiang, Q.; Huo, Z. Highly selective hydrothermal production of cyclohexanol from biomass-derived cyclohexanone over Cu powder. *Catal. Today* **2016**, *274*, 94–98. [CrossRef]
5. Sikhwivhilu, L.M.; Coville, N.J.; Naresh, D.; Chary, K.V.; Vishwanathan, V. Nanotubular titanate supported palladium catalysts: The influence of structure and morphology on phenol hydrogenation activity. *Appl. Catal. A Gen.* **2007**, *324*, 52–61. [CrossRef]
6. Wang, L.; Zhang, Y.; Du, R.; Yuan, H.; Wang, Y.; Yao, J.; Li, H. Selective one-step aerobic oxidation of cyclohexane to ε-caprolactone mediated by N-hydroxyphthalimide (NHPI). *ChemCatChem* **2019**, *11*, 2260–2264. [CrossRef]
7. Alazman, A.; Belic, D.; Alotaibi, A.; Kozhevnikova, E.F.; Kozhevnikov, I.V. Isomerization of cyclohexane over bifunctional Pt-, Au-, and PtAu-heteropoly acid Catalysts. *ACS Catal.* **2019**, *9*, 5063–5073. [CrossRef]
8. Kim, A.R.; Ahn, S.; Yoon, T.U.; Notestein, J.M.; Farha, O.K.; Bae, Y.S. Fast cyclohexane oxidation under mild reaction conditions through a controlled creation of Redox-active Fe (II/III) Sites in a Metal-organic Framework. *ChemCatChem* **2019**, *11*, 5650–5656. [CrossRef]
9. Chary, K.V.; Naresh, D.; Vishwanathan, V.; Sadakane, M.; Ueda, W. Vapour phase hydrogenation of phenol over Pd/C catalysts: A relationship between dispersion, metal area and hydrogenation activity. *Catal. Commun.* **2007**, *8*, 471–477. [CrossRef]
10. Nelson, N.C.; Manzano, J.S.; Sadow, A.D.; Overbury, S.H.; Slowing, I.I. Selective hydrogenation of phenol catalyzed by palladium on high-surface-area ceria at room temperature and ambient pressure. *ACS Catal.* **2015**, *5*, 2051–2061. [CrossRef]
11. Zhang, X.; Du, Y.; Jiang, H.; Liu, Y.; Chen, R. Insights into the stability of Pd/CN catalyst in liquid phase hydrogenation of phenol to cyclohexanone: Role of solvent. *Catal. Lett.* **2019**, *149*, 1–10. [CrossRef]
12. Liu, T.; Zhou, H.; Han, B.; Gu, Y.; Li, S.; Zheng, J.; Zhong, X.; Zhuang, G.L.; Wang, J.G. Enhanced selectivity of phenol hydrogenation in low-pressure CO_2 over supported Pd catalysts. *ACS Sustain. Chem. Eng.* **2017**, *5*, 11628–11636. [CrossRef]
13. Sanyal, U.; Song, Y.; Singh, N.; Fulton, J.L.; Herranz, J.; Jentys, A.; Gutiérrez, O.Y.; Lercher, J.A. Structure sensitivity in hydrogenation reactions on Pt/C in aqueous-phase. *ChemCatChem* **2019**, *11*, 575–582. [CrossRef]
14. Singh, N.; Lee, M.S.; Akhade, S.A.; Cheng, G.; Camaioni, D.M.; Gutiérrez, O.Y.; Glezakou, V.A.; Rousseau, R.; Lercher, J.A.; Campbell, C.T. Impact of pH on aqueous-phase phenol hydrogenation catalyzed by carbon-supported Pt and Rh. *ACS Catal.* **2018**, *9*, 1120–1128. [CrossRef]
15. Wang, L.; Zhang, J.; Yi, X.; Zheng, A.; Deng, F.; Chen, C.; Ji, Y.; Liu, F.; Meng, X.; Xiao, F.S. Mesoporous ZSM-5 zeolite-supported Ru nanoparticles as highly efficient catalysts for upgrading phenolic biomolecules. *ACS Catal.* **2015**, *5*, 2727–2734. [CrossRef]
16. Nakagawa, Y.; Ishikawa, M.; Tamura, M.; Tomishige, K. Selective production of cyclohexanol and methanol from guaiacol over Ru catalyst combined with MgO. *Green Chem.* **2014**, *16*, 2197–2203. [CrossRef]
17. Martinez-Espinar, F.; Blondeau, P.; Nolis, P.; Chaudret, B.; Claver, C.; Castillón, S.; Godard, C. NHC-stabilised Rh nanoparticles: Surface study and application in the catalytic hydrogenation of aromatic substrates. *J. Catal.* **2017**, *354*, 113–127. [CrossRef]
18. Kinoshita, A.; Nakanishi, K.; Yagi, R.; Tanaka, A.; Hashimoto, K.; Kominami, H. Hydrogen-free ring hydrogenation of phenol to cyclohexanol over a rhodium-loaded titanium (IV) oxide photocatalyst. *Appl. Catal. A Gen.* **2019**, *578*, 83–88. [CrossRef]
19. Wang, Y.; Yao, J.; Li, H.; Su, D.; Antonietti, M. Highly selective hydrogenation of phenol and derivatives over a Pd@carbon nitride catalyst in aqueous media. *J. Am. Chem. Soc.* **2011**, *133*, 2362–2365. [CrossRef]

20. Zhao, C.; Song, W.; Lercher, J.A. Aqueous phase hydroalkylation and hydrodeoxygenation of phenol by dual functional catalysts comprised of Pd/C and H/La-BEA. *ACS Catal.* **2012**, *2*, 2714–2723. [CrossRef]
21. Li, Z.; Liu, J.; Xia, C.; Li, F. Nitrogen-functionalized ordered mesoporous carbons as multifunctional supports of ultrasmall Pd nanoparticles for hydrogenation of phenol. *ACS Catal.* **2013**, *3*, 2440–2448. [CrossRef]
22. Galkin, M.V.; Sawadjoon, S.; Rohde, V.; Dawange, M.; Samec, J.S. Mild Heterogeneous Palladium-Catalyzed Cleavage of β-O-4′-Ether Linkages of Lignin Model Compounds and Native Lignin in Air. *ChemCatChem* **2014**, *6*, 179–184. [CrossRef]
23. Paone, E.; Espro, C.; Pietropaolo, R.; Mauriello, F. Selective arene production from transfer hydrogenolysis of benzyl phenyl ether promoted by a co-precipitated Pd/Fe$_3$O$_4$ catalyst. *Catal. Sci. Technol.* **2016**, *6*, 7937–7941. [CrossRef]
24. Wu, H.; Song, J.; Xie, C.; Wu, C.; Chen, C.; Han, B. Efficient and mild transfer hydrogenolytic cleavage of aromatic ether bonds in lignin-derived compounds over Ru/C. *ACS Sustain. Chem. Eng.* **2018**, *6*, 2872–2877. [CrossRef]
25. Zhang, Q.; Li, H.; Gao, P.; Wang, L. PVP-NiB amorphous catalyst for selective hydrogenation of phenol and its derivatives. *Chin. J. Catal.* **2014**, *35*, 1793–1799. [CrossRef]
26. Gilkey, M.J.; Xu, B. Heterogeneous transfer hydrogenation as an effective pathway in biomass upgrading. *ACS Catal.* **2016**, *6*, 1420–1436. [CrossRef]
27. Croston, M.; Langston, J.; Sangoi, R.; Santhanam, K.S.V. Catalytic oxidation of p-toluidine at multiwalled functionalized carbon nanotubes. *Int. J. Nanosci.* **2002**, *1*, 277–283. [CrossRef]
28. Croston, M.; Langston, J.; Takacs, G.; Morrill, T.C.; Miri, M.; Santhanam, K.S.V.; Ajayan, P. Conversion of aniline to azobenzene at functionalized carbon nanotubes: A possible case of a nanodimensional reaction. *Int. J. Nanosci.* **2002**, *1*, 285–293. [CrossRef]
29. Gordon, M.; Santhanam, K.S.V. High product yield in a narrow column configuration of carbon nanotubes: A Pathway for nanosynthetic machine. *Proc. Electrochem. Soc.* **2003**, *13*, 285–288.
30. Zhou, M.; Tian, L.; Niu, L.; Li, C.; Xiao, G.; Xiao, R. Upgrading of liquid fuel from fast pyrolysis of biomass over modified Ni/CNT catalysts. *Fuel Process. Technol.* **2014**, *126*, 12–18. [CrossRef]
31. Zhou, M.; Zhu, H.; Niu, L.; Xiao, G.; Xiao, R. Catalytic hydroprocessing of furfural to cyclopentanol over Ni/CNTs catalysts: Model reaction for upgrading of bio-oil. *Catal. Lett.* **2014**, *144*, 235–241. [CrossRef]
32. Zhang, J.; Chen, C.; Yan, W.; Duan, F.; Zhang, B.; Gao, Z.; Qin, Y. Ni nanoparticles supported on CNTs with excellent activity produced by atomic layer deposition for hydrogen generation from the hydrolysis of ammonia borane. *Catal. Sci. Technol.* **2016**, *6*, 2112–2119. [CrossRef]
33. Li, G.; Han, J.; Wang, H.; Zhu, X.; Ge, Q. Role of dissociation of phenol in its selective hydrogenation on Pt (111) and Pd (111). *ACS Catal.* **2015**, *5*, 2009–2016. [CrossRef]
34. Song, Y.; Chia, S.H.; Sanyal, U.; Gutiérrez, O.Y.; Lercher, J.A. Integrated catalytic and electrocatalytic conversion of substituted phenols and diaryl ethers. *J. Catal.* **2016**, *344*, 263–272. [CrossRef]

© 2020 by the authors. Licensee MDPI, Basel, Switzerland. This article is an open access article distributed under the terms and conditions of the Creative Commons Attribution (CC BY) license (http://creativecommons.org/licenses/by/4.0/).

Review

Supercritical Carbon Dioxide Extraction of Lignocellulosic Bio-Oils: The Potential of Fuel Upgrading and Chemical Recovery

Nikolaos Montesantos and Marco Maschietti *

Department of Chemistry and Bioscience, Aalborg University, Niels Bohrs Vej 8A, 6700 Esbjerg, Denmark; nmo@bio.aau.dk
* Correspondence: marco@bio.aau.dk

Received: 31 January 2020; Accepted: 30 March 2020; Published: 1 April 2020

Abstract: Bio-oils derived from the thermochemical processing of lignocellulosic biomass are recognized as a promising platform for sustainable biofuels and chemicals. While significant advances have been achieved with regard to the production of bio-oils by hydrothermal liquefaction and pyrolysis, the need for improving their physicochemical properties (fuel upgrading) or for recovering valuable chemicals is currently shifting the research focus towards downstream separation and chemical upgrading. The separation of lignocellulosic bio-oils using supercritical carbon dioxide (sCO$_2$) as a solvent is a promising environmentally benign process that can play a key role in the design of innovative processes for their valorization. In the last decade, fundamental research has provided knowledge on supercritical extraction of bio-oils. This review provides an update on the progress of the research in sCO$_2$ separation of lignocellulosic bio-oils, together with a critical interpretation of the observed effects of the extraction conditions on the process yields and the quality of the obtained products. The review also covers high-pressure phase equilibria data reported in the literature for systems comprising sCO$_2$ and key bio-oil components, which are fundamental for process design. The perspective of the supercritical process for the fractionation of lignocellulosic bio-oils is discussed and the knowledge gaps for future research are highlighted.

Keywords: lignocellulosic; bio-oil; biocrude; upgrading; supercritical extraction; supercritical CO$_2$; hydrotreatment; biorefinery; pyrolysis; hydrothermal liquefaction

1. Introduction

The contemporary society is heavily dependent on fossil fuels, both for energy and for the production of chemicals and materials. Indicatively, in 2017, about 10^8 barrels/day of crude oil, 10^{10} m^3/day of natural gas, and 22 Mt/day of coal were consumed worldwide [1,2]. The volumetric figures for crude oil and natural gas correspond to approximately 11 Mt/day and 12 Mt/day, respectively. The CO$_2$ emissions related to the consumption of fossil fuels amounted to 96 Mt/day in 2017. In addition, the use of crude oil, natural gas, and coal is predicted to increase by 20%, 32%, and 10% by 2050 [1]. These figures clearly show that the development of efficient technological pathways for substantially increasing the production share of energy, chemicals, and materials from biomass, in partial substitution of fossil fuels, is a key aspect for reducing net CO$_2$ emissions and paving the way for a sustainable society based on renewable resources.

Biomass can be classified into first and second generation. First generation biomass is considered edible biomass, which can be extensively cultivated expressly for energy production [3]. Related technological examples are the production of bioethanol from corn and sugar cane and the production of biodiesel from soybean [4]. Second generation biomass is non-edible biomass, characterized by lignocellulosic structure and typically available in the form of waste or by-products from forestry, agriculture, municipal waste management, and the pulp and paper industry [3]. Not being in direct competition with food production, with respect to land and water utilization, the development of process pathways for exploiting second generation biomass is particularly appealing. Low-value utilization of residual lignocellulosic biomass as a source of heating by direct combustion is widespread, both in households (e.g., wood pellets) and in industrial plants where it is produced (e.g., lignin from pulp and paper industry). On the other hand, higher value-added utilizations (i.e., production of liquid fuels and chemicals) are still extremely limited. An indicative example is that lignocellulosic biofuels from forestry and agricultural residues (i.e., bio-oils) account for only 3% of worldwide biofuel production [5].

In line with the above, a huge research effort has been made in the last few decades, aimed at advancing the technologies for the conversion of lignocellulosic feedstocks into liquid fuels and chemicals. Thermochemical processes have shown promising results, as in the case of pyrolysis and hydrothermal liquefaction (HTL). The main product of these processes is a lignocellulosic crude bio-oil, along with gas and biochar as side-products [6]. In the majority of scientific literature, the lignocellulosic oil produced by means of pyrolysis is named bio-oil, whereas the lignocellulosic oil produced by means of HTL is named biocrude (or bio-crude). The same approach is followed in this review. However, in this work, the term bio-oil is also used when pyrolysis and HTL oils are discussed jointly.

With regard to fuel production, crude bio-oils require both physical and chemical upgrading steps in order to be inserted into the current technological chain (e.g., blending with specific petroleum fuels). The utilization of typical refinery processes developed for fossil fuels is challenging, owing to the marked difference between bio-oils and petroleum. More specifically, from a chemical standpoint, lignocellulosic bio-oils (LC bio-oils) suffer from high average molecular weights (MWs), high oxygen and water content, as well as high acidity. From a physical standpoint, they suffer from high density and viscosity [7,8]. The production of specific chemicals is also challenging, as LC bio-oils are complex mixtures of a large number of compounds that show diverse molecular weight and polarity. The diversity in polarity is a marked difference compared with fossil crudes, which makes existing separation and chemical processes developed for fossil crudes not straightforwardly applicable to LC bio-oils. On the one hand, the development of thermochemical conversion processes of lignocellulosic biomass has been the focus in previous years; on the other hand, it is expected that the research focus in the coming years will shift towards the development of separation processes to be applied downstream of the thermochemical conversion unit. Such processes aim to produce either upgraded fuel fractions or specific value-added chemicals.

Among separation processes available in the chemical industry for the fractionation of oils, distillation and liquid–liquid extraction (LLE) are particularly relevant and widespread. However, in the case of LC bio-oils, the high molecular weight leads to very high distillation temperatures or very high vacuum requirements. For example, distillate fractions not exceeding 50 wt%–60 wt% of the bio-oil are reported for pressures as low as 0.1 mbar [9–12]. In addition, temperatures above 100 °C during distillation promote side-reactions (e.g., polymerization) [8,13]. Another factor negatively affecting the distillation of bio-oils is their water content. It was reported to reduce the efficiency of the separation and to lead to unsteady boiling and process control difficulties [9]. In addition, the water content is one of the main factors reducing the viscosity of bio-oils [8]. This implies that, in the lower stages of a continuous-flow distillation column, the bio-oil flowing downwards is expected to be extremely viscous, leading to operational problems. With regard to LLE, the process requires large quantities of organic solvents, the majority of which are produced from petroleum (e.g., dichloromethane, n-pentane) [14,15], thus spoiling one of the selling points of renewable fuels by using petroleum-based materials for bio-oil processing. In addition, organic solvents are often noxious and their recovery downstream of the extraction requires an additional process step (e.g., solvent evaporation).

An alternative for the separation of oils is represented by the extraction using supercritical carbon dioxide (sCO$_2$) as a solvent. In the supercritical region (i.e., above 73.8 bar and 31 °C), carbon dioxide behaves as a liquid solvent, exhibiting liquid-like densities and good solvent power towards apolar and moderately polar compounds. In addition, sCO$_2$ has favorable transport properties (e.g., high diffusivity, low viscosity), which make it an efficient solvent. Downstream of the separation unit, carbon dioxide can be easily recovered by, for example, partial decompression, and recycled. From a solvent perspective, CO$_2$ is environmentally benign, safe (non-flammable), low-cost, and readily available. It is considered a particularly valid alternative for the separation of oils with high boiling temperatures (i.e., low volatility) [16]. For example, it finds industrial application for the fractionation of low volatility liquid mixtures, as in the case of hydroxyl-terminated perfluoropolyether oligomers [17]. The supercritical process is expected to have operational advantages as the dissolution of sCO$_2$ in the liquid phase causes oil expansion and a drop in viscosity, which facilitate the flow of the oil in continuous countercurrent equipment. As an example, these mechanisms are exploited in enhanced oil recovery processes based on CO$_2$ injection, which are particularly effective with respect to heavy oils and tar sands [18].

For the reasons stated above, research papers reporting the use of sCO$_2$ for separating LC bio-oils have appeared in the literature in the last decade. The main objectives of this review paper are as follows: (i) to provide an update on the progress of the research on sCO$_2$ separation of LC bio-oils; (ii) to provide an update on the progress of the research on high-pressure phase equilibria of systems comprising sCO$_2$ and bio-oil components; and (iii) to highlight knowledge gaps inspiring future research work in this area. The review paper is structured as follows. Section 2 describes the properties of LC bio-oils, also in relation to the starting biomass and the thermochemical conversion process, and highlights the issues encountered in the downstream upgrading aimed at fuel and chemicals production. Section 3 provides basic features of sCO$_2$ extraction processes and reviews in detail the literature providing experimental data on sCO$_2$ extraction of LC bio-oils. Section 4 analyzes available experimental data of phase equilibrium of carbon dioxide at supercritical conditions and key components of LC bio-oils and provides data correlations and interpretation based on the Chrastil model. Section 5 summarizes the authors' view with regard to the integration of this technology in the downstream upgrading of LC bio-oils and highlights research and technology gaps.

2. Lignocellulosic Bio-Oils

2.1. Lignocellulosic Feedstocks

Second generation lignocellulosic biomass is abundant in the form of agricultural, forestry, and municipal residues, as well as industrial byproducts such as lignin from the pulp and paper industry. In 2017, the agriculture sector was estimated to be able to generate from 11 to 47 Mt/day of lignocellulosic residues, whereas forestry residues were estimated to be 2.1 Mt/day [19]. Their quantitative potential as raw materials alternative to fossil fuels is thus significant. However, when considering the inherent difficulties in collecting a sparse resource and conveying it to conversion plants, together with the yields of transformation into valuable products, it is probable that only a fraction of fossil-based fuels and chemicals can realistically be substituted by LC counterparts. The potential contribution of municipal sewage sludge is rather small (e.g., approximately 25 kt/day of dry biomass in the European Union (EU) in 2010 [20]), even though it is worth considering it in the context of an overall effort aimed at raw material shift from fossil fuels to renewables. With regard to industrial lignocellulosic residues, Kraft lignin from the pulp and paper industry and lignin-rich residues from bio-ethanol plants (i.e., residual enzymatic lignin) are currently made available in small quantities, with an estimated 0.19 Mt/day [21] and 0.74–2.2 kt/day [22], respectively. In spite of the small quantities currently available, lignin is attractive owing to the peculiar chemical structure, which is composed of aromatic moieties. In addition, lignin is available with reproducible quality as a by-product of industrial systems that are either well-established (Kraft process) or under development (lignocellulosic to ethanol). These aspects make lignin an interesting by-product for the production of fuel additives, as well as bulk and fine aromatic chemicals [23,24]. Another interesting industrial example is represented by residues of the palm industry, such as palm kernel shells. In 2006, Malaysia produced approximately 0.14 Mt/day of lignocellulosic residues associated to palm oil production, which can be an attractive feedstock for the production of bio-oils [25].

Lignocellulose consists in its majority of three natural macromolecules (i.e., cellulose, hemicellulose, and lignin), as well as a small weight percent of ash (i.e., inorganics). The ratio between these macromolecules varies widely from biomass to biomass and is one of the parameters affecting the composition of bio-oils. Key examples of these feedstocks that were studied in the literature, with respect to their conversion to bio-oils, are reported in Table 1, with typical ranges of mass fraction for cellulose, hemicellulose, lignin, and ash.

Table 1. Lignocellulosic biomass and distribution of cellulose, hemicellulose, lignin, and ash (wt%) on a water-free basis [26].

Residual Biomass Type	Forestry		Agricultural		Industrial	Municipal
	Poplar	Pine	Sugarcane Bagasse	Wheat Straw	Kraft Lignin	Sewage Sludge [1]
Cellulose	41–49	38–50	34–42	29–52	0–1	-
Hemicellulose	17–33	18–30	19–43	11–39	0–1	-
Lignin	18–32	23–28	19–21	8–30	90	-
Ash	0–2	0–6	2–12	1–14	1–2	26–55

[1] Organic content is reported as total volatile matter in the range of 40 wt%–74 wt%.

Besides the composition in terms of cellulose, hemicellulose, and lignin, the overall elemental composition is another basic parameter that affects the properties of the bio-oil that can be obtained by LC biomass. Table 2 reports the oxygen content, the H/C and O/C ratios, and the higher heating value (HHV) for specific LC feedstocks that were studied in the context of bio-oil production. Elemental data and HHV are given on a water-free basis. HHV is calculated as reported in the literature [27], when the experimental data were not provided.

Table 2. Examples of lignocellulosic biomass, studied for bio-oil production, and their properties. Elemental composition, ash, and higher heating value (HHV) are reported on a water-free basis. Water content is also reported when available.

Biomass Type	Industrial Residue	Softwood	Hardwood	Energy Crops	Agricultural Residues
	Kraft Lignin	Pine Bark	Beech	Wheat Stalk	Sugarcane Bagasse
Water (wt%)	32.6	NA	8.7	10.5	NA
Oxygen (wt%)	26	42.13	44.5	47.9	52.5
H/C	1.04	1.38	1.38	1.53	2.00
O/C	0.31	0.64	0.69	0.79	1.00
Ash	0.8	1.07	0.8	NA	NA
HHV (MJ/kg)	27.67	20.2[1]	19.2[1]	17.8[1]	16.4[1]
Reference	[28]	[29]	[30]	[31]	[32]

[1] Calculated; NA: not reported.

As can be seen, the elemental composition of woody biomass is rather constant for pine (softwood) and beech (hardwood). Lignin is the biomass with the lowest oxygen content, which results in the highest HHV. The crop residues have the highest oxygen content, and thus the lowest HHVs.

2.2. Lignocellulosic Bio-Oils

Bio-oils are defined here as the organic-rich liquid product of the thermochemical conversion of biomass. The two most prominent conversion processes are pyrolysis and hydrothermal liquefaction (Table 3). Pyrolysis employs high temperature to thermally break the macromolecules constituting the biomass in an oxygen-free environment. Drying of the biomass is required prior to pyrolysis. Depending on the residence time, the process is denoted as fast (i.e., a few seconds) or slow (i.e., hours to days). Microwaves can be used as an alternative heating source [8]. HTL can handle both dry and wet biomass, whose macromolecular constituents are broken down by a complex set of reactions in subcritical or supercritical water environment, with or without assisting chemicals (e.g., catalysts, pH regulators, co-solvents) [33].

Table 3. Thermochemical conversion methods for bio-oil production [8,33]. HTL, hydrothermal liquefaction.

Process	Pretreatment	Temperature	Pressure	Bio-Oil Yield
Pyrolysis	Drying, size reduction	500–600 °C	Atmospheric	up to 75 wt%
HTL	Size reduction	250–450 °C	100–350 bar	up to 75 wt%

Pyrolysis has reached industrial production level, with several plants around the world [8]. Licensed pyrolysis technologies (e.g., BTG-BTL, Ensyn, VTT) are being used in plants that produce bio-oil mostly from forestry residues. Characteristic examples are the Empyro plant (Twence – Empyro) in the Netherlands with production of approximately 65 t/day [34] and the Côte-Nord plant (Ensyn) in Canada with a capacity of approximately 130 t/day [35]. The HTL technology is utilized in several pilot plants around the globe [33] and one demonstration plant is under construction in Norway by Steeper energy and Silva Green Fuel with a production capacity of approximately 4 t/day [36].

LC bio-oils produced by pyrolysis and HTL are typically viscous dark liquids (Figure 1), composed to a large extent of oxygenated organic components. These oils are tight water-in-oil emulsions with water mass fractions typically in the range of 20 wt% to 30 wt% for pyrolysis oils [37], while lower values are observed for HTL biocrudes (4 wt%–15 wt%) [38–42]. Owing to the polarity induced by oxygen to many chemical constituents, raw bio-oils are not fully miscible with hydrocarbon solvents. They are, however, miscible with oxygen-containing organic solvents such as acetone and tetrahydrofuran [40,43]. In some cases, inorganics (i.e., ash) are present in bio-oils. They can either originate from the biomass or be introduced during processing [44]. Some quantitative information concerning physical and chemical properties of bio-oils is available in the literature and is reviewed in the following.

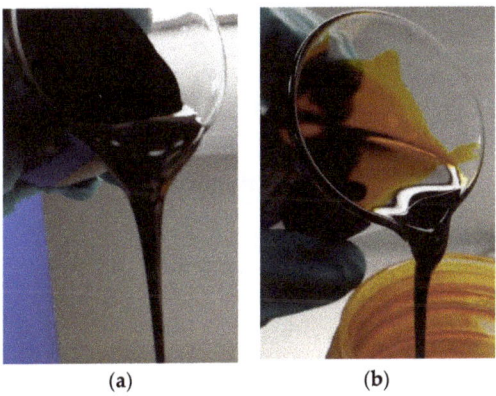

Figure 1. Example of lignocellulosic bio-oil from pinewood. (**a**) hydrothermal liquefaction (HTL); (**b**) pyrolysis.

Density values for LC bio-oils are typically higher than 1000 kg/m^3, with small variations depending on the source of biomass. For example, values between 970 kg/m^3 and 1100 kg/m^3 are typical for HTL biocrudes [9,45,46]. Somewhat higher values are typically reported for pyrolysis oils, namely between 1100 kg/m^3 and 1200 kg/m^3 [47]. Density values are thus higher than petroleum and petroleum liquid products, which typically range from 800 kg/m^3 for light crude oils up to 1000 kg/m^3 for heavy oils and bitumens [48].

Kinematic viscosity values for LC pyrolysis oils are reported in a broad range, from 7 to 53 cSt at 40 °C [37,49]. The variation is strongly connected to the water content of the bio-oil (i.e., 17 wt%–48 wt%), with viscosity markedly decreasing with the water content. In another work, the kinematic viscosity is reported to be 28 cSt at 60 °C [50]. With regard to LC biocrudes (obtained by means of HTL), dynamic viscosity values are reported in a much broader range, namely from 1700 cP to 4·10^6 cP [30,51,52]. The higher values correspond to semisolids and result from the drying of the biocrude. Another work reports the kinematic viscosity of a dehydrated HTL biocrude being 12 cSt at 40 °C [9]. The major difference in viscosity between pyrolysis and HTL oils is, at least in large part, owing to the difference in the water content. As reported data typically refer to different water contents, caution is required in drawing conclusions related to the quality of the bio-oil on the basis of viscosity.

With regard to bulk chemical properties, the most important characteristic of LC bio-oils is the high oxygen content (O), which can widely range from as low as 10 wt% to as high as 50 wt%. In Table 4, a few examples are reported. As can be seen, oxygen ranges from 19 wt% to 40 wt%. In all cases, the oxygen of the bio-oil is lower than that of the biomass, which also translates into a higher HHV and a lower O/C ratio. Nevertheless, the oxygen values are at least one order of magnitude higher than the typical values for crude oils, where they lie in the range 0.05 wt% to 1.5 wt% [53]. Therefore, deoxygenation of LC bio-oil is a requirement in the perspective of fuel production.

Table 4. Properties of HTL bio-oils for different biomass feedstocks. Oxygen (O), H/C, O/C, ash, and higher heating value (HHV) on a water-free basis.

Biomass Type	Industrial Residue		Softwood	Hardwood		Energy Crop	Agricultural Residue	
	Kraft Lignin	Palm Shell	Pine Bark	Beech	Eucalyptus	Wheat Stalk	Wheat Straw	Sugarcane Bagasse
Process	HTL	Pyrolysis	HTL	HTL	Pyrolysis	HTL	Pyrolysis	HTL
Oxygen (wt%)	21	33	28.3	27.3	23.9	18.8	40.0	36.3
H/C	1.11	1.74	1.2	1.19	1.08	1.29	1.35	1.64
O/C	0.23	0.43	0.33	0.30	0.26	0.20	0.56	0.49
HHV (MJ/kg)	31.7	27	27.4[1]	28.3[1]	29.2	32.4[1]	21.9	24.8[1]
Reference	[28]	[54]	[29]	[30]	[55]	[31]	[56]	[32]

[1] Calculated as reported in the literature [27].

The inorganic content of bio-oils, cumulatively reported as ash, is the result of both the presence of metals in the original biomass and their introduction during processing. Examples of metals found in bio-oils are sodium (Na), potassium (K), and iron (Fe) [44,57,58]. Pyrolysis oils typically have a low ash content (e.g., 0.01 wt%–0.2 wt%) [47,59], as most of the metals are not contained in volatile compounds, and thus do not transfer in the gas stream. In addition, entrained particles in the gas stream are retained by filters. The small amounts of ash reported in pyrolysis oils are typically associated to volatile organometallic components [44]. On the other hand, HTL biocrudes are typically obtained by processes where catalysts (e.g., potassium carbonate [57]) and pH regulators (e.g., sodium hydroxide [60]) are utilized. These chemicals dissolve in the water droplets emulsified in the biocrude, resulting in high ash contents. For example, when an alkali catalyst is used, the ash content of the biocrude can be as high as 5 wt% [45]. With regard to the biomass feedstock, Anastasakis et al. [41] performed non-catalytic HTL of miscanthus and spirulina, which contained 2.7 wt% and 6.5 wt% of ash, respectively, and the biocrudes ended up on average with 2.8 wt% and 6.6 wt% ash, respectively. It is important to note that even metal content values as low as 0.5 wt% can be detrimental for the downstream catalytic upgrading (e.g., hydrotreating). Therefore, the metal content of bio-oils must be substantially reduced if the oil is to be hydrotreated [57].

Another relevant characteristic of LC bio-oils is the acidity. The total acid number (TAN) of LC bio-oils is reported in the range of 9 to 200 mg KOH/g [30,59–64], depending on the bio-oil and on the measurement method. TAN is a representation of the acidity of the liquid mixture and is a cumulative effect of carboxylic acids and phenolic components that are present in the LC bio-oils. Considering these two partial acid numbers, namely the carboxylic acid number (CAN) and phenolic acid number (PhAN), the determination of the latter is not always achieved, when methods developed for petroleum are used. In some cases, the reported (cumulative) TAN values are essentially CAN. A modification of ASTM D664 [65] was reported by Christensen et al. [63], which successfully determines both acid numbers in pyrolysis oils. Montesantos et al. [64] measured the TAN of a LC biocrude from HTL, using a method inspired by this modification, and reported a CAN value of 43 mg KOH/g and TAN of 129 mg KOH/g. On the other hand, most of the literature for HTL biocrudes reports values of TAN up to 67 mg KOH/g [30,61,66]. This suggests that, at least in some cases, CAN values are those actually reported. Therefore, the methodology for the determination of TAN for bio-oils is one of the properties that requires standardization to ensure meaningful comparisons between different works. In this respect, Oasmaa et al. [27] published an interesting review of properties and analytical methods for the case of LC pyrolysis oils.

With regard to the detailed chemical structure of LC bio-oils, they are complex mixtures consisting of an overwhelming number of chemical components. Ketones, phenols, organic acids, and aromatic hydrocarbons are commonly found in the volatile fraction [7,67] of LC bio-oils. Other components such as aldehydes, esters, furans, and sugars are reported in the volatile fraction of pyrolysis oils [67]. The nonvolatile fraction of bio-oils contains mainly oligomers with several carbon and oxygen atoms (e.g., 15–29 carbon and 8–10 oxygen atoms [68]). This heavy fraction includes both phenolic and carbonyl functional groups [68,69], but little is known in detail. The average molecular weight of bio-oils is typically in the range of 300 to 1000 g/mol [70–72], with individual components with molecular weight ranging from less than 100 g/mol to several thousand g/mol [15,68,73].

More information is available for the volatile fraction, which is typically studied by gas-chromatography coupled with mass spectrometry (GC-MS). A discussion on this fraction is reported in the following. In most cases, the relative amounts of components of the volatile fraction are simply reported in terms of chromatographic peak area ratios. When internal standards are used, the mass fraction of identified components rarely accounts for more than 50 wt% of the bio-oil. The highest values are typically associated with pyrolysis oils, owing to several low molecular weight components (e.g., acetol, acetic acid, and glycoladehyde) that each constitute up to 10 wt% of the oil. In addition, levoglucosan is typically found in woody pyrolysis oils at high mass fractions (e.g., 10 wt%) [43,54,74–78]. On the other hand, single components at such a high concentration are not observed in the volatile fraction of HTL biocrudes, which is characterized by total mass fractions of identified components in a lower range (e.g., 10 wt%–30 wt%) [28,40,64,76,79,80]. However, chemical classes like polyaromatic hydrocarbons (e.g., retene) [57] and long chain fatty acids (e.g., hexadecanoic acid) [64] were found to constitute up to 9 wt% and 4 wt% [58] of HTL biocrudes, respectively. Such components are often not reported in the characterization of HTL biocrudes, even though they seemingly are a considerable part of it.

Table 5 reports the main chemical classes observed in the volatile fraction of LC bio-oils, together with typical ranges of molecular weight and number of carbon atoms (carbon number). In addition, an example is provided in which two pinewood bio-oils, produced by fast pyrolysis and HTL, are directly compared. The chemical classes include components with a wide range of molecular weights (i.e., 60 g/mol to above 300 g/mol) and volatilities, with boiling points ranging from around 100 °C (as normal boiling points) to values by far exceeding 300 °C (as atmospheric equivalent temperature, AET). The presence of these components in LC bio-oils results in high oxygen content, polarity, and acidity.

Table 5. Typical chemical classes identified by gas-chromatography coupled with mass spectrometry (GC-MS) in bio-oils with examples of mass fractions in pyrolysis oils and HTL biocrudes from pinewood and examples of specific components [10,43,57,58,76,81–84]. MW, molecular weight.

Chemical Class	Pyrolysis [43]	HTL [58]	MW	Carbon Number	Examples
Ketones	Up to 8 wt%	Up to 0.5 wt%	74–124	C3–C10	Hydroxyacetone Cyclopenten-1-one, 2-Cyclopentanone, 2,5-dimethyl
Phenols	Up to 0.1 wt%	Up to 0.3 wt%	94–122	C6–C8	Phenol o-Cresol 4-Ethylphenol
Guaiacols	Up to 0.5 wt%	Up to 0.6 wt%	124–178	C7–C10	Guaiacol Eugenol Creosol
Benzenediols	-	Up to 1.7 wt%	110–124	C6–C8	Catechol 4-Ethylcatechol
Short chain fatty acids[1]	Up to 5 wt%	Up to 0.2 wt%	60–144	C2–C8	Acetic acid Octanoic acid
Long chain fatty acids	-	Up to 3.8 wt%	172–284	C10–C19	Decanoic acid Octadecanoic acid
Aromatic acids	-	Up to 1.8 wt%	152–300	C8–C20	Dehydroabietic acid Benzeneacetic acid, 3-hydroxy
Furans	Up to 0.5 wt%	-	84–132	C4–C8	Furfural Furanone, 2(5H)-
Aldehydes	Up to 8 wt%	-	60–152	C2–C8	Glycolaldehyde Benzaldehyde, 3-hydroxy-4-methyl-
Esters	-	-	130–296	C6–C19	Benzoic acid, 4-methoxy-, methyl ester Furoic acid methylester
Sugars	Up to 10 wt%	-	132–144	C5–C6	Levoglucosan 2,3-Anhydro-d-galactosan
Benzenes	-	Up to 1 wt%	92–134	C7–C10	o-Cymene Toluene
Polyaromatic hydrocarbons	-	Up to 9 wt%	128–234	C10–C18	Naphthalene Retene

[1] For simplicity, small carboxylic acids (i.e., acetic, propanoic) are included in this class.

As different lignocellulosic biomasses own different fractions of cellulose, hemicellulose, and lignin, it follows that some component types will be favored during thermochemical conversion. More specifically, a larger fraction of lignin increases the fraction of phenolic components such as phenol, alkylphenols, guaiacols, and benzenediols [85]. For example, the mass fraction of phenolic components (relative to the total mass fraction of GC-MS identified components) reported by Belkheiri et al. [40] for an HTL biocrude from Kraft lignin was 97%, whereas the analogous ratio (in terms of peak areas) observed by Pedersen et al. [57] for an HTL biocrude from aspen wood was only 27%. Other chemical classes like ketones, furans, and acids are abundant in bio-oils that originate from biomass with a high content of cellulose and hemicellulose [86]. For example, the peak area fraction of ketones reported by Pedersen et al. [57] was 21% for a biocrude originating from biomass with 67% of cellulose and hemicellulose. The analogous quantity reported by Chan et al. [87] was 21% in a biocrude originating from biomass with 50% of cellulose and hemicellulose.

Another important aspect of LC bio-oils is the stability under storage. Kosinkova et al. [46] reported an increase in density of about 5% for an HTL biocrude under ambient conditions upon 25 weeks of storage. The increase reached 30% when the biocrude was stored at 43 °C for the same duration. The density increase was connected to the increase of the average molecular weight, which in turn resulted from polymerization reactions of certain lignin-derived phenolic components. Nguyen et al. [79] observed composition changes in a lignin-derived HTL biocrude (lignin oil), which can be attributed to instability of the biocrude. Specifically, alkylphenols, benzenediols, and phenolic dimers decreased over time, while the average molecular weight of the oil increased after two years under ambient conditions. The increase in MW was observed by means of gel permeation chromatography (GPC), and was in line with the decrease of the GC-MS identified fraction of components from 15 wt%, for the fresh biocrude, to 11 wt% after long-term storage. Interestingly, a diethyl ether extracted fraction (corresponding to 66 wt% of the biocrude) was very stable, with the identified GC-MS fraction exhibiting only a 1% reduction in two years at ambient conditions. An important observation of this work is that the presence of inorganic solids in the lignin oil catalyzes the polymerization reactions, resulting in a higher aging rate. Elliott et al. [37] reported an increase of the kinematic viscosity of a fast pyrolysis oil between 60% and 70% during an aging test at 80 °C for 24 h. This viscosity difference, together with the reported increase of the average molecular weight, is indicative of the relatively low stability of pyrolysis oil. A similar increase of viscosity was observed after 12 months at 21 °C. Storage at 5 °C and −17 °C resulted in smaller increases of viscosity, equal to 19% and 7%, respectively [37].

2.3. Valorization

2.3.1. Fuel Upgrading

Hydrotreating (HT) is the most prominent process for removing heteroatoms. It is a catalytic process adopted from the mature oil industry, where it is typically performed at temperatures of 90–390 °C and pressures of 15–170 bar. HT aims to remove heteroatoms like sulfur, nitrogen, oxygen, and hydrogenate C = C bonds (hydrogenation, HYD). The main expenditure is the consumption of hydrogen (H_2), with typical refinery values in the range of 10 to 850 Nm^3 of H_2 per m^3 of feed. The variation in the process conditions depends on the feedstock composition as well as on the objective of HT. Higher heteroatom contents usually require more severe conditions. Typically, fixed bed reactors are used, with the most common catalysts being cobalt-molybdenum (CoMo) and nickel molybdenum (NiMo) [88].

With regard to bio-oils, HT was studied at laboratory scale, with the main objective being the hydrodeoxygenation (HDO) of the liquid feed [89,90]. Reaction temperatures and pressures are reported in the range of 150 to 400 °C and 40 to 140 bar, with most typical values being 300–400 °C and 100 bar [39,89–91]. The most commonly used catalysts are commercially available CoMo and NiMo on Al_2O_3 support, although several other catalysts were also studied [92]. The economics of the process is mainly affected by the high H_2 partial pressures required. In particular, the large oxygen fraction of bio-oils requires H_2 to feed ratios in the range of 300 to 600 Nm^3/m^3 [9,89,93], which are similar to the ratios required for the sulfur-rich heavy fractions of crude oil refineries (e.g., residual oil) [88].

One of the issues of HT is the formation of coke, which leads to gradual deactivation of the catalyst. Typically, high molecular weight and high boiling point components accelerate deactivation because of deposition on the catalyst active sites [94]. Bjelic et al. [15] investigated the chemistry of a wood-derived HTL biocrude, as well as the chemistry of the extract and the residue obtained from the hydrotreated biocrude by means of liquid–liquid extraction using n-pentane (C5). The presence of HT resistant species in the residue was observed and the recommendation of separating the C5-insoluble fraction from the biocrude prior to HT was formulated. The metal content of some bio-oils (mainly HTL biocrudes) is also expected to pose problems to HT, because of rapid deactivation of the catalyst. Metal deposition is irreversible; it substantially reduces the catalytic activity and increases the pressure drop in the HT reactor [88,94]. When metals are deposited on the catalyst bed, regeneration (e.g., to remove coke) can sinter the catalyst surface, resulting in area loss [94]. These factors make necessary the demetallization of bio-oils prior to HT.

In spite of the abovementioned problems, research on HT of bio-oils has shown promising results indicating that, using optimal catalysts and conditions for hydrotreating, fuel grade oils can be achieved. Jensen et al. [90] performed HT experiments on HTL lignocellulosic biocrude on a commercial $NiMo/Al_2O_3$ catalyst. This parametric study highlighted the importance of high temperature and high H_2 partial pressure for achieving high-levels of HDO. The maximum HDO level attained in this work corresponded to a reduction of oxygen from 5.3 wt% (feed biocrude) down to 0.1 wt% (hydrotreated oil), with operating conditions of 350 °C, 97 bar, and an H_2 to biocrude ratio of 500 Nm^3/m^3. The same authors performed GCxGC-MS on HTL biocrude and its HT product, and qualitatively reported (i.e., based on peak area ratios) an increase of alkanes and cycloalkanes from about 10%–15% to more than 50%. Another important observation of this work regards the total acid number, which was reduced to zero after hydrotreating. These results prove the feasibility of HT of LC bio-oils at laboratory scale. Nevertheless, further work is needed on a larger scale to verify the process operability and economics, especially with respect to catalyst deactivation rates and fouling. The physical upgrading (separation) of LC bio-oils upstream of the HT unit may be a key factor to improve the HT operability on commercial catalysts on an industrial scale.

2.3.2. Production of Green Chemicals

Another perspective of valorization of LC bio-oils is the production of green chemicals. Pyrolysis oils contain some chemicals at relatively high mass fractions. Among them are acetic acid (up to 9 wt%) and acetol (up 8 wt%) [84]. Acetol is used as intermediate to produce polyols and acrolein [95], and acetic acid is an important chemical with a production exceeding 33 kt/day. Other chemicals with high mass fractions are glycolaldehyde (up to 6 wt%) and levoglucosan (up to 9 wt%) [68]. Even though it has no industrial application at this moment, levoglucosan has been identified as a potential building block for the chemical synthesis of high value-added pharmaceutical products [96–99]. Phenol is one of the most studied chemicals in LC bio-oils owing to its huge global demand, which has reached 27 kt/day in 2015 [100]. Phenol and its derivatives (e.g., guaiacol) can be used to produce resins and adhesives, as well as in the pharmaceuticals, food, or perfumery industries. Phenol mass fractions in LC bio-oils typically range from 0.1 wt% to 2 wt% [101], with values up to 5 wt% reported in the literature for pyrolysis oils from high-lignin content bio-mass [54]. High-value specialty chemicals are also found in LC bio-oils, albeit in low mass fractions. For example, vanillin, which is produced in majority by petroleum-derived guaiacol (approximately 85% of world production) [102], can be found in pyrolysis oils between 0.1 wt% and 1 wt% [43,54,84,101,103]. The production of vanillin in 2018 reached 100 t/day [104].

Recently, the antioxidant activity of bio-oil fractions has been studied for both pyrolysis and HTL oils derived by different biomasses [105–108]. Phenolic dimers and oligomers are suggested as the active antioxidant components, as monomers exhibited small to no antioxidant activity. The phenolic fractions were compared with commercial stabilizing agents (i.e., butylated hydroxytoluene, BHT) and showed identical or even better antioxidant activity in bio-diesel and bio-lubricants [106,107].

3. sCO$_2$ Separation of Bio-Oils

3.1. sCO$_2$ Basics

Carbon dioxide exists in a supercritical state at conditions that exceed 73.8 bar and 31 °C. The pressure–temperature (P–T) phase diagram of pure CO_2, plotted from experimental data available in the literature [109–111], is shown in Figure 2. Even though CO_2 is a low-density vapor at standard conditions (i.e., 0 °C and 1 bar), in the supercritical region, it can exhibit liquid-like densities while keeping relatively high diffusivities and low viscosities [112]. The presence of high-density regions, at pressures not exceedingly high, allow sCO$_2$ to exhibit solvent power comparable to liquid solvents in pressure ranges where separation processes are feasible.

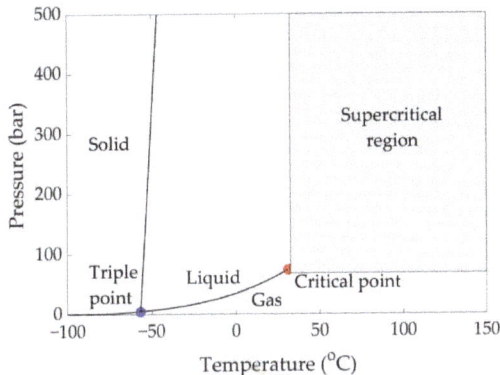

Figure 2. CO_2 phase diagram. Data taken from the literature [109–111].

The density of CO_2 at the critical point is approximately 468 kg/m^3, while its viscosity is approximately 0.03 cP [109]. Within the supercritical region, the density varies widely depending on pressure and temperature. For example, for temperatures and pressures in the range of 35 to 150 °C and 75 to 400 bar, respectively, it varies between 105 kg/m^3 and 973 kg/m^3. Even at the highest densities, sCO_2 exhibits viscosities remarkably lower than those of typical liquid solvents, such as hexane (i.e., approximately four times lower). In addition, in a broad area of the supercritical region, the density of sCO_2 can be varied remarkably with relatively small variations of pressure and temperature. This aspect provides a high tunability of the solvent characteristics, namely solvent power and selectivity, on the basis of two degrees of freedom (i.e., pressure and temperature). This is a distinct advantage over conventional liquid solvents, where a single parameter (i.e., temperature) can be varied to alter the properties of the solvent.

Another interesting property of CO_2 is that, even though it has zero dipole moment [113], it has a large quadrupole moment, which makes it a good solvent for both apolar and low polarity compounds [114]. In addition, sCO_2 is non-toxic, not flammable, and widely available at low cost. Its use in renewable chemical production is likely to be neutral with respect to CO_2 emissions into the atmosphere, as CO_2 does not need to be produced on purpose, contrary to most petroleum-derived organic solvents. Moreover, it may even be speculated that a spread in utilization of sCO_2 in industrial applications has the potential of a slight reduction of CO_2 emissions, as its storages will increase in the growth period of the sCO_2 technology. Furthermore, underground CO_2 storage facilities might be utilized in combination with units employing sCO_2 as a solvent for renewable production processes, thus taking advantage of the in situ presence of high-pressure CO_2 and developing negative CO_2-emission processes.

Typical process unit configurations include semi-continuous (i.e., batch) extraction, which can be performed either in a single stage extractor or a multi-stage column, and countercurrent continuous extraction [16,112,113,115]. In semi-continuous single-stage extractions (Figure 3a), the feed material is charged in a high-pressure extractor vessel and sCO_2 is continuously delivered to the bottom of the extractor. The CO_2-rich stream exits the vessel from the top and is expanded in a separator in order to release the extracted matter by reduction of the solubility, while the solute-free solvent is recompressed and recirculated. Alternatively, membranes or adsorbents can be used to separate the solutes from the sCO_2 without depressurization. Such an example is the use of activated carbon in the supercritical decaffeination of coffee beans [116]. In the case of multi-stage semi-continuous operation (Figure 3b), a reflux loop is added, where part of the extract is refluxed at the top of the column. In this mode of operation, the feed is contacted with the ascending CO_2-rich phase in a multiple-stage manner to achieve a better separation, compared with the single-stage operation [117]. The unextracted material (i.e., raffinate) remains in the vessel until the end of the batch extraction.

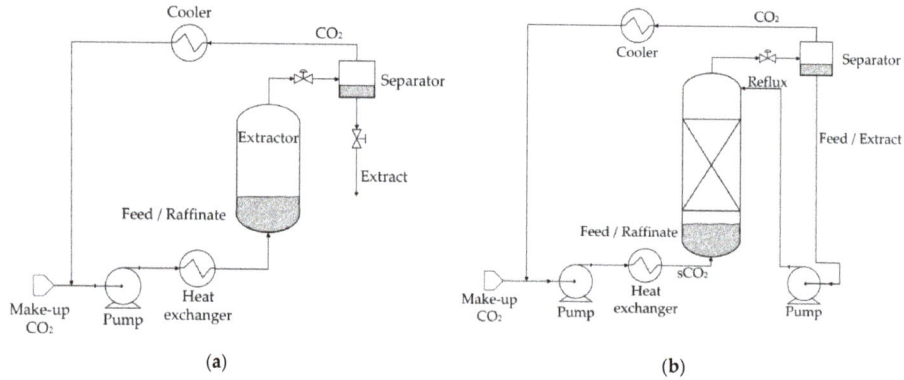

Figure 3. Flow diagram of semi-continuous extraction processes. (**a**) Single stage; (**b**) multi-stage.

The continuous countercurrent operation involves continuous flow of the feed from an entry point at the top or at an intermediate point of an extraction column, while the sCO$_2$ flows continuously from the bottom. Depending on the feed entry point, the extract is either continuously collected (Figure 4a) or partly recompressed and refluxed at the top of the column (Figure 4b). In both cases, the raffinate exits from the bottom. Another mode of operation for the reflux process is the use of a temperature gradient in the top section of the column (the enrichment section), which serves to induce a drop in solubility, producing an internal reflux. This mode of operation can be exploited in some sCO$_2$-oil systems, depending on the P–T region, and is based on retrograde condensation phenomena [16,118].

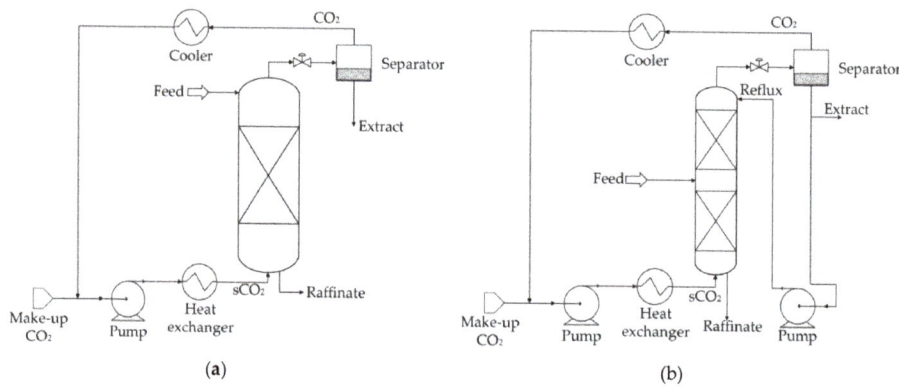

Figure 4. Flow diagram of continuous countercurrent extraction process. (**a**) Without reflux; (**b**) with reflux.

A few industrial applications have been established to exploit the advantages of sCO$_2$ as a solvent, which are mostly extractions from solid matter. Examples are coffee and tea decaffeination and the extraction of essential oils from plant feedstocks [113]. Such extractions from solid material are typically performed as batch semi-continuous operations by intermittently charging batches of the solid (e.g., coffee beans) and removing part of the spent materials from the bottom of the system without depressurizing (i.e., without shut down), while adding fresh material from the top [113]. A niche industrial application of sCO$_2$ on liquid feeds is the fractionation of perfluoropolyethers, aimed at narrowing the molecular weight distribution of polymer fractions used as lubricants [17]. In addition, separation of many other oils proved feasible using sCO$_2$ as a solvent. Such cases are the deterpenation of citrus oils [117,119], separation of fish oil ethyl esters [118,120,121], extraction of squalene from vegetable oils [16], and purification of frying oil [122,123]. These research works show the potential of continuous countercurrent systems for oily feeds. Literature studies on sCO$_2$ extraction applications of crude oil mainly suggest the use of sCO$_2$ for fractionation of heavy oils and bitumens, aimed at recovering a lighter extract, separating it from a heavy asphaltenic residue. The extracts show lower molecular weight, boiling point, and viscosity, thus being more easily conveyed to other refinery units, while the residue can be used for electricity production [124,125].

3.2. sCO$_2$ Extraction of Lignocelulosic Bio-Oils

The majority of the literature studies on sCO$_2$ separation of LC bio-oils refer to pyrolysis oils, with only a few cases referring to HTL biocrudes. Table 6 summarizes these works and their main aspects, regarding the nature of the biomass feedstock, the thermochemical conversion process producing the bio-oil, the system size, and the year of publication. All data so far are limited to laboratory-scale sCO$_2$ extraction systems and the semi-continuous single-stage mode of operation, with the exception of Mudraboyina et al. [126], where the extractor was coupled with a rectification column with a temperature gradient, allowing an internal reflux operation. The data selection was delimited to literature referring to extraction of LC bio-oils obtained by phase separation (e.g., gravity settling) of the reaction products of the thermochemical process. In particular, this means that extraction of LC bio-oil species dissolved or dispersed in water or in organic solvents is not considered relevant in this context.

Table 6. Experimental studies of semi-continuous sCO$_2$ fractionation of bio-oils.

Feedstock	Thermochemical Process	Extractor Volume (cm^3)	Year	Ref.
Pine	HTL	178	2020	[58]
Pine	HTL	178	2019	[127]
Pine	HTL	178	2019	[64]
Palm kernel shell	Slow pyrolysis	50	2018	[128]
Pine	Fast pyrolysis	640	2017	[43]
Palm kernel shell	Slow pyrolysis	50	2017	[129]
Beech	Slow pyrolysis	640	2016	[130]
Red pine	Fast pyrolysis	25	2016	[131]
Kraft lignin	Microwave pyrolysis	160[1]	2015	[126]
Beech	Slow pyrolysis and fast pyrolysis	600	2015	[84]
Sugarcane bagasse and cashew shells	Pyrolysis	-	2011	[132]
Wheat-hemlock	Fast pyrolysis	-	2010	[103]
Wheat-sawdust	Fast pyrolysis	-	2009	[133]

[1] The extractor was coupled with a rectification column.

A generalized laboratory scale system is shown in Figure 5, which represents all literature studies except for Mudraboyina et al. [126], where the extractor was coupled with a rectification column. Such a typical system utilizes a CO$_2$ cylinder for supplying the solvent. CO$_2$ is subcooled via a heat exchanger and pumped as liquid to pressurize the vessel. CO$_2$ can be supplied to the system by different types of positive displacement pumps such as pneumatic [64], syringe [126], and diaphragm [43]. In some cases, the pumped CO$_2$ is preheated before entering the extractor [43,84,126,128–130]. The extraction vessel may contain an insert that can be dismounted for easy charging of the feed and retrieving the residue [64,127].

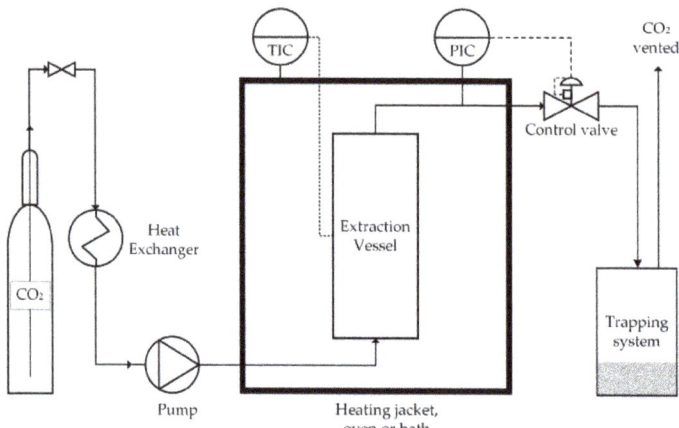

Figure 5. Generalized semi-continuous sCO$_2$ extraction system.

An inert packing material can be used to provide feed dispersion, thus increasing the contact area of CO$_2$ and the feed. For this purpose, glass beads [64,103,127,128,133] can be used, or a porous media (e.g., silica, activated carbon) [43,84,130,131] on which the bio-oil can be adsorbed prior to the sCO$_2$ extraction. In the latter case, some authors reported the introduction of effects that render the interpretation of the extraction results more complex. More specifically, Feng and Meier [43] reported different extraction yields between activated carbon and silica at the same extraction conditions, with the former achieving lower extraction yield than the latter at most extraction conditions, whereas the effect was inverted at 300 bar and 80 °C. However, the difference in these results is not adequate to provide a clear understanding of the effect of the two adsorbents. The potential of using experimental data that combine extraction and adsorption effects is limited in the perspective of scaling up the process to large-scale production, as the adsorbent use would raise issues for continuous operation. Exceptions would be presented only for batch separations aimed at extracting small quantities of high-value compounds, where solid–liquid interactions can be advantageously exploited.

With regard to practical aspects, the use of a non-return valve [64] at the bottom of the extraction vessel is advantageous in order to avoid back-flow of liquid feed and fouling of the upstream pipeline. This is particularly important for foulant residues that cannot be dissolved by pure sCO$_2$ and are difficult to dissolve with most solvents, as is the case for LC bio-oils. The extraction vessel can be heated by a heating jacket [64], an oven [128], or a bath [126], in order to maintain a constant extraction temperature. An automatic or manual valve can be used downstream the extractor to control the flow rate of CO$_2$. Downstream of the valve, the expanded fluid releases the components that are dissolved in sCO$_2$, which are collected in a trapping system while CO$_2$ gas is vented. This type of system is adequate for research purposes, although a potential industrial application would require separation of CO$_2$ from the extracts and a solvent recycle loop.

The trapping system should be considered carefully and ensure adequate collection of the extracts. In this regard, it is particularly important to (i) ensure a low temperature to allow maximum condensation of the extracts dissolved in the CO$_2$-rich phase (the gas phase); (ii) limit entrainment of liquid droplets and solid particles in the CO$_2$ gas stream exiting the expansion valve; and (iii) ensure a smooth flow of viscous extracts downstream of the expansion valve until the collection container. Such a trapping system can consist of a series of elements (e.g., washing bottles) that serve to capture the condensed extract, followed by additional elements such as adsorbents (e.g., activated carbon) or absorbents (e.g., water) to reduce entrainment of fine droplets. The trap can be at ambient or freezing temperature. The latter can greatly improve the condensation as well as reduce evaporation of the collected condensed extracts exposed to CO$_2$ flow during long extractions. This can be essential for pyrolysis oils that contain a large mass fraction of relatively low-boiling organic components (e.g., acetic acid) and water. The latest work of Feng and Meier [43] highlights this important detail, where the authors used several traps at ambient conditions, including cotton wool, activated carbon, and water, while reporting mass balance discrepancies at the range of 6 wt% to 14 wt%. The same range of losses was encountered during the extraction of HTL lignocellulosic bio-oil, albeit using a cold (approximately 5) trapping system, but no adsorbent [64,127]. Another consideration is the viscosity of the extract, which increases as the extraction progresses. This may lead to the requirement of heating the tube downstream of the expansion valve in order to improve the flow towards the trap.

3.3. Experimental Conditions, Extraction Yields, and Vapor Phase Loadings

Table 7 reports the operating parameters (i.e., extraction pressure (P) and temperature (T), initial mass of the feed (F), CO_2 mass flow rate, and solvent-to-feed ratio (S/F)) and the outcome of the experiments of sCO_2 extraction of LC bio-oils that were found in the literature. The outcome is reported in terms of total extraction yield (Y) and vapor phase loading (VPL). Y is the ratio of the total mass of extract to the mass of the feed. VPL is defined as the mass of extract retrieved for a given mass of sCO_2 that flowed in the extraction vessel (i.e., the solute loaded in the solvent). It is typically expressed as g of extract per kg of sCO_2 in a specific time interval. In addition, the density of pure sCO_2 is reported, taken from the NIST database [109].

Table 7. Experimental conditions and technical data of experiments of sCO_2 extraction of lignocellulosic bio-oils found in the literature. Pressure (P), temperature (T), CO_2 density, feed mass (F), CO_2 mass flow rate, solvent-to-feed ratio (S/F), total extraction yield (Y), and vapor phase loading (VPL) are reported.

P (bar)	T (°C)	CO_2 Density (kg/m^3)	F (g)	CO_2 Flow (g/min)	S/F (g/g)	Y (wt%)	VPL (g/kg)	Ref.
330–450	80–150	531–851	49–54	4.7–5.9	30.0–36.7	44.1–53.4	5.9–99.7	[58]
247–448	120	500–730	28.0–30.8	4.8–6.9	40.6–50.9	33.9–48.9	2.7–46.5	[127]
112–400	40–120	548–882	40.9–51.1	3.2–7.0	12.8–85.5	17.1–41.8	13.1–36.5	[64]
300–400	50–70	788–923	1.8–2.4	3.1–3.8	78.7–116.3	4.7–12.0	0.4–1.0	[128]
100–300	60–80	221–830	100^1	8.3^1	30^1	0.1–14.3	0.03–4.8	[43]
150–400	33–66	691–961	2.5	1.1–8.3	26.6–199.0	4.2–30.4	0.9–2.2	[129]
200	60	723	40–80	8.3	37.5–75.0	23.4–40.9	3.1–10.9	[130]
100–300	50	384–870	1.0	0.4–0.9	56.3	71.1^2	30.6^3	[131]
80–100	35 (45–95)^4	490–700	2–5	2–10	40–100	11–31	2.2–5.1	[126]
150–250	60	603–786	80	10	45	7.3–41.4	1.6–11.5	[84]
120–300	50	510–870	100	11.7–20	21–36	9–15	4.2–4.3	[132]
100–300	40	628–910	50	40	288	46	0.7–2.16	[103]
250–300	45	857–890	50	30	288	45	0.7–2.8	[133]

^{1} Normalized data are provided only; ^{2} achieved with the use of co-solvent (i.e., methanol); ^{3} average value; ^{4} rectification column temperatures in parentheses.

The temperature in the literature studies was in the range of 40 to 120 °C. The pressure was largely varied between 80 bar and 448 bar, with the aim of a wide range of solvent density (i.e., 221 to 961 kg/m^3). However, only the studies of Chan et al. [128] and Montesantos et al. [58,64,127] explored pressures higher than 300 bar. CO_2 flowrates used in these studies ranged from values much lower than 1 g/min up to 40 g/min, which, in some cases, resulted in very low extraction yields or very high solvent-to-feed ratios. Depending on the applied conditions, extraction yield values in the range 0.1 wt% to 48.9 wt% are reported using pure sCO_2. Cheng et al. [131] achieved up to 71 wt% extraction yield with the use of up to 25 vol% of methanol as a co-solvent.

Regarding the effect of the extraction conditions, some important conclusions can be drawn. The increase of pressure at constant temperature is consistently reported as beneficial for extraction yields and vapor phase loadings. This is straightforwardly explained by the increase of the solvent power (i.e., increased solubility of bio-oil components in sCO_2). The effect of temperature increase at constant pressure is more complex, as it leads to improved mass transfer, but it can also lead to reduced solubility. Feng and Meier [43], on pyrolysis oils, and Montesantos et al. [58,64], on HTL biocrudes, observed that a temperature increase at constant pressure generally increases Y and VPL, especially at high pressures (e.g., 300–400 bar). This indicates that, even though the increase of temperature is often associated with reduced solvent power, because of the decrease of solvent density, the enhanced mass transfer parameters improve the extraction rate and can improve Y and VPL of the process. In addition, Montesantos et al. [64] reported operational problems (i.e., sporadic clogging of the system) with HTL bio-oils at low temperatures. These problems were severe at 40 °C and moderate at 60 °C, while smooth operation was achieved at 80 °C and above.

3.4. Physical and Chemical Properties of sCO$_2$ Extracts

sCO$_2$ extracts exhibit lower viscosity and density compared with the bio-oil feed. Patel et al. [132] reported the kinematic viscosity of sugarcane bagasse pyrolysis oil sCO$_2$ extracts and compared it with previously published viscosities of similar non-fractionated LC bio-oils [50]. The extracts, which accounted for only 9 wt%–15 wt% of the feed, exhibited a lower viscosity (i.e., approximately 18 cSt) than the crude bio-oil (approximately 28 cSt). It is important to note that, in an aging test of 60 days, the extracted oil showed greater stability with regard to viscosity, with an increase of only 4 cSt compared with approximately 22 cSt for the non-fractionated LC bio-oil.

Wang et al. [134] performed extraction of a corn stalk pyrolysis oil by a sequence of pressurization–depressurization steps (intermittent extraction), and reported that the density of the bio-oil (i.e., 1150 kg/m^3) was reduced down to 952–980 kg/m^3 for the extracts. Montesantos et al. [64] reported a reduction from 1051 kg/m^3 down to 941–1017 kg/m^3 for the extracts of HTL bio-oil from pinewood. Although the density reduction is moderate, from a fuel perspective, the sCO$_2$ extracts exhibit densities in line with residual marine fuels, which are typically in the range of 920 to 1010 kg/m^3 [135].

In the work of Montesantos et al. [64], TAN reductions from 129 mg KOH/g down to 61 to 120 mg KOH/g are reported for the sCO$_2$ extracts, with the TAN values increasing with the extraction progression (i.e., over time). Carboxylic acids were sequentially extracted, with short chain fatty acids extracted at the earlier stages and long chain fatty acids at later stages of the extraction. In addition, the acidity of sCO$_2$ residue shifted from carboxylic to phenolic nature, as the fatty acids were extracted preferentially with respect to more polar and heavier phenolic components.

Regarding water content, Feng and Meier [84] reported a reduction in both the extract and the residues for two fast pyrolysis oils. Even though this result clearly indicates the operational difficulties in collecting the extracted water in the trap, and thus the difficulties in closing the water mass balances, it also indicates that water is co-extracted. Therefore, sCO$_2$ extraction can be used as a means for dewatering the LC bio-oil while fractionating it. In the work of Feng and Meier [84], the water content of the two feeds was 19 wt% and 43 wt%; for the extract, it ranged from 10 wt% to 13 wt% and 14 wt% to 19 wt%; and for the residues, it ranged from 6 wt% to 8 wt% and 10 wt% to 13 wt%, respectively. Reduction of water in the sCO$_2$ extract of a HTL biocrude was observed by Montesantos et al. [58] as well, where the extracts were dewatered up to 77%, with respect to the initial water content.

Regarding metal content, in a recent work, Montesantos et al. [58] focused on the effect of sCO$_2$ extraction on the removal of metals from HTL biocrude. The authors reported 95%–98% removal of metals (i.e., metal content reduced from 8500 mg/kg to lower than 200 mg/kg), which represents a significant improvement of the quality of the extracts, with respect to the feed biocrude, in light of the downstream hydrotreating. With regard to the elemental composition, the difference between sCO$_2$ extracts and their respective feeds can be seen in Figure 6, where the H/C atomic ratio and the oxygen mass fractions on a water-free basis are compared. In the case of extracts, values corresponding to extractions at different pressure and temperature conditions on the same feed were averaged.

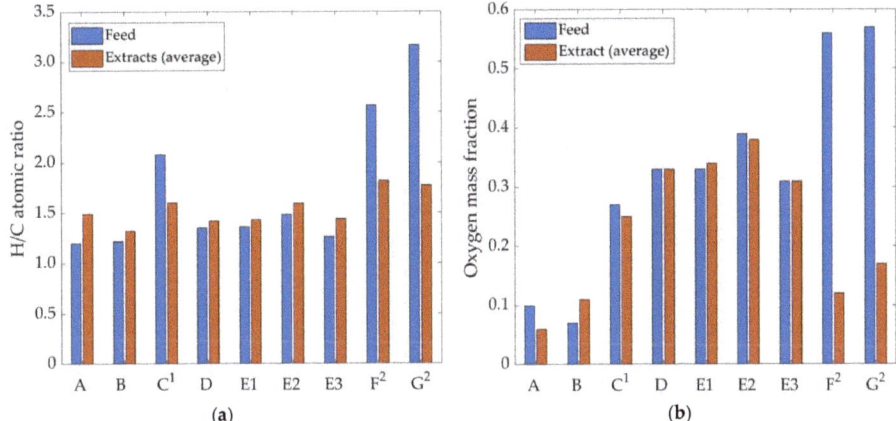

Figure 6. (a) H/C atomic ratio and (b) oxygen mass fraction of bio-oils (feed) and their sCO$_2$ extracts on a water-free basis. A: Montesantos et al. [58]; B: Montesantos et al. [64]; C: Chan et al. [129]; D: Feng and Meier [130]; E1–E3: Feng and Meier [84]; F: Naik et al. [103]; G: Rout et al. [133]. E1–E3 correspond to different feeds. [1] It was assumed that the data were reported on a water-free basis (unspecified by the authors); [2] it was assumed that the data were reported including water (unspecified by the authors).

In most cases, the H/C ratio and oxygen mass fraction are not affected, or are moderately affected, by the extraction. In particular, the absence of significant variation of the oxygen content indicates that oxygen is widespread in many classes of molecules of different polarity and molecular weight, making it difficult to have a selective process in this regard. For an HTL biocrude, Montesantos et al. [58] reported a partial deoxygenation (i.e., oxygen mass fraction 0.10 down to 0.05–0.07). The oxygen reduction can be explained in terms of the increased VPL of alkylbenzenes and polyaromatic hydrocarbons, especially in the first extracts, in the presence of water in the biocrude (extraction of nondewatered bio-oil). The data from Naik et al. [103] and Rout et al. [133] appear to be strong outliers. In these works, a fast pyrolysis oil with an extremely high oxygen content is reported, which is higher than the oxygen content of the biomass feedstock. In addition, the majority of GC-MS identified components are oxygenated organic molecules, with a lower oxygen content than the oxygen reported by elemental analysis. All things considered, the data from Naik et al. [103] and Rout et al. [133] seem to show some inconsistencies.

3.5. Chemical Composition sCO$_2$ Extracts

The components identified in the volatile fraction of sCO$_2$ extracts of LC bio-oils can be generally grouped into the following chemical classes: (1) ketones; (2) phenols; (3) guaiacols; (4) benzenediols; (5) aldehydes; (6) esters; (7) furans; (8) syringols; (9) short chain fatty acids (SFAs); (10) long chain fatty acids (LFAs); (11) aromatic acids (ArAcid); (12) single-ring aromatic hydrocarbons (benzenes); and (13) polyaromatic hydrocarbons (PAHs). These classes, with indication of the key components reported in the literature, are presented in Table 8.

Table 8. Typical chemical classes and key components observed in sCO$_2$ extracts [43,58,64,84,103,126, 129,130,132,133].

Number	Class	MW Range	Key Components	Carbon Number
1	Ketones	74–124	Acetol	3
			Acetylacetone	5
			2-pentanone	5
			Propan-2-one. 1-acetyloxy-	5
			2-Cyclopenten-1-one, 2,3-dimethyl-	7
2	Phenols	94–122	Phenol	6
			o-cresol	7
			m-cresol	7
			p-cresol	7
			2,5-Dimethylphenol (p-xylenol)	8
3	Guaiacols	124–178	Guaiacol	7
			4-methyl guaiacol	8
			4-ethyl guaiacol	8
			4-propyl guaiacol	8
			Eugenol	10
			Isoeugenol	10
			Creosol	8
			Vanillin	8
4	Benzenediols	110–124	1,2-Benzenediol	6
5	Aldehydes	60–152	Glycolaldehyde	2
6	Esters	130–296	Pentanoic acid. 4-oxo-. methylester	6
7	Furans	84–126	Furfural	5
			5-Hydroxymethylfurfural	6
8	Syringols	154	Syringol	8
9	Short chain fatty acids	60–144	Acetic acid	2
			Propionic acid	3
			Hexanoic acid	6
10	Long chain fatty acids	256–284	n-Hexadecanoic acid	16
			n-Octadecanoic acid	18
11	Aromatic acids	136	Benzeneacetic acid, 3-hydroxy	8
12	Benzenes	120	o-Cumene	10
13	Polyaromatic hydrocarbons	128	Naphthalene	10

Most of these classes include low molecular weight components. Some chemical classes comprise components characterized by a narrow molecular weight range, whereas fatty acids [64] and esters [84], observed in HTL and pyrolysis oil extracts, respectively, are characterized by a wide range of molecular weight (see Table 5). The boiling point distribution in the extracts is in a wide range as well, from as low as 117 °C for acetic acid, reaching up to 400 °C for the high molecular weight fatty acids.

Most components are typically in low amounts in the sCO$_2$ extracts (below 1 wt%) of both oils, with a few exceptions. In extracts of pyrolysis oils, short chain carboxylic acids, acetol, and glycolaldehyde are observed in large amounts; in some cases, as high as 20 wt% [43,84,130]. In addition, guaiacol, furfural, and syringol are reported with noteworthy mass fractions (1 wt%–5 wt%) in extracts of LC bio-oils of both pyrolysis and HTL. As a result of the concentration of these components, the sCO$_2$ extracts exhibit a larger volatile fraction, with the fraction of the oil that can be identified by GC-MS being larger than the feed (e.g., up to 70 wt% [84]). On the other hand, levoglucosan (a sugar) present in pyrolysis oils (see Section 2.2.) is not extracted and remains entirely in the residue.

In an attempt of highlighting trends and providing some quantitative indication, values of distribution factors (K-values) were calculated from the available data reported in the literature. The calculated K-values are reported in Figure 7. using a box plot, covering the range of P, T, and composition corresponding to the experimental conditions. It should be noted that, as compositional information mostly refers to the lighter fraction of the bio-oil, most of the identified components are extracted preferentially, thus with K-values higher than 1.

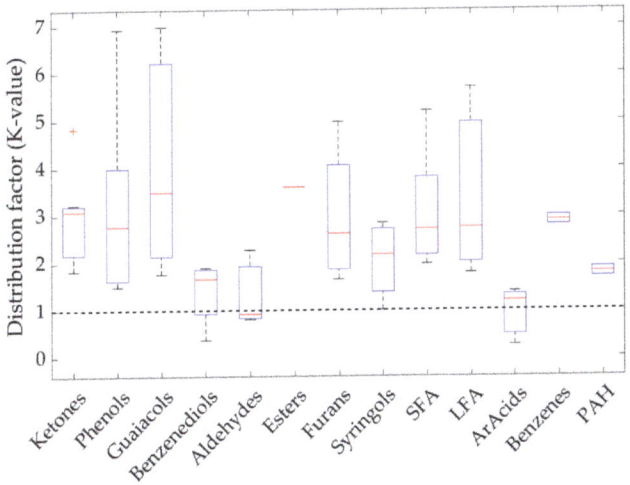

Figure 7. Distribution factors (K-values) of component classes identified in the literature. K-values correspond to the pressure, temperature, and compositional ranges reported in the literature. The boxes include the 50% distributed around the average, the red line is the median value, the whiskers are the extreme values (over 75% and under 25%), and the "+" denotes values considered outliers [58,64,84,126, 127,129,130]. SFA, short chain fatty acid; LFA, long chain fatty acid; PAH, polyaromatic hydrocarbon.

As can be seen, guaiacols are among the most preferentially extracted components, with the highest median K-value (close to 4). Among other hydroxybenzenes, phenols, syringol, and benzenediols show descending K-values in the range of 2.8 down to 1.8. This is in line with the solubility predictions performed by Maqbool et al. [101], which show that guaiacols have the highest solubility in sCO_2, followed by phenols and benzenediols. The lower K-values of benzenediols are connected with their higher polarity, owing to the presence of two hydroxyl groups. They were, however, observed to be extracted with yields up to 60 wt%, after the preferentially extracted components (e.g., guaiacols) were depleted in the feed [64]. Low molecular weight ketones and furans (see Table 8) are other classes of components preferentially extracted, with K-values close to 3. The extracted components of these classes show similar molecular weight and polarity. Glycolaldehyde is the only aldehyde found in a considerable amount in pyrolysis oils (up to 8 wt%), and has one of the lowest MWs of the reported components [43]. For this component, the available literature data do not allow calculating K-values, although it was reported by Feng and Meier [43] to be remarkably concentrated up to 16 wt%. Esters are typically not detected in the feeds, but low molecular weight esters (i.e., 100–130) were reported in a single case in the sCO_2 extracts (up to 1.6 wt%), indicating high expected K-values [84].

With regard to the acids, most literature works report acetic and propionic acids, which are found in pyrolysis oil feed at high concentrations. Their K-values are typically above 2–3, which is in line with their low molecular weight. Montesantos et al. [58,64,127] identified several fatty acids ranging between C5 and C18, as well as two aromatic acids. As an example, Figure 8 shows the K-values of said acids in terms of extraction progression (extraction time increasing from left to right). All reported extractions were at 120 °C, but with different pressures (and thus different solvent densities). As can be seen, the short chain fatty acids (i.e., C5–C8, SFA) maintain high K-values for the whole duration of the extraction, which is in line with their relatively low MW. On the contrary, the long chain fatty acids (i.e., C6–C18, LFA) and the aromatic acids (i.e., dehydroabietic acid, ArAcid) initially exhibit K-values lower than 1 and, at the latest stages, they are extracted with K-values up to 8 and 4, respectively. This figure highlights the possibility of sequential extraction and fractionation of different acids on the basis of their molecular weight. From a theoretical standpoint, it also highlights the high dependency of K-values on the composition of the mixture, which reflects the strong thermodynamic non-ideality of the system sCO_2 + bio-oil.

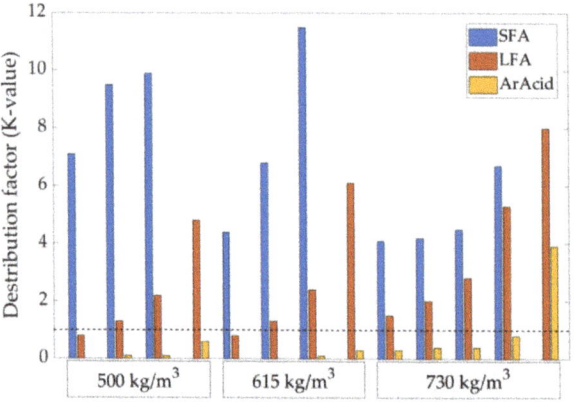

Figure 8. Variation of K-values with extraction progression (left to right) of different types of acids: short chain fatty acids (SFAs); long chain fatty acids (LFAs); aromatic acids (ArAcid). Extraction temperature: 120 °C. Three different solvent densities [127].

Finally, the only cases reporting the extraction of aromatic hydrocarbons are the works of Montesantos et al. [58,127] on an HTL biocrude. In these works, alkylbenzenes (e.g., o-cymene) exhibited the highest K-values (up to 6) and were extracted with yields up to 90 wt%. Polyaromatic hydrocarbons were the largest fraction of the extract (up to 23 wt%) and included some of the highest molecular weight components (e.g., retene). They exhibited increasing K-values with extraction progression, as the lower MW components were depleted from the feed, achieving extraction yields up to 83 wt%. The possibility of extracting these components is in line with the known capability of sCO_2 of extracting hydrocarbons, even at a very high molecular weight.

4. High Pressure Solubility Data and Modelling

The process design of sCO_2 extraction and downstream separation of LC bio-oils requires the development of suitable thermodynamic models (e.g., based on cubic equations of state) in order to estimate the K-values of the different species to be separated, for a given pressure, temperature, and overall composition of the system. In turn, the development of such models for complex mixtures as LC bio-oils requires strategies of lumping bio-oil components into classes, defining thermodynamic properties (e.g., critical properties and acentric factors) for each class of components, as well as collecting experimental data of high-pressure phase equilibrium of sCO_2 and selected class-representative components. Experimental data are used for model validation and model tuning by estimation of optimal values of binary interaction parameters. This procedure is analogous to what has been developed for decades for petroleum reservoir fluids, also in relation to sCO_2 injection studies [136]. However, crude LC bio-oils (i.e., prior to hydrotreating) are drastically different from fossil crude oils, with a large amount of oxygen and polar components spread in the entire molecular weight range, which leads to notes of complexity in the definition of optimal lumping strategies.

In addition, literature data on high-pressure phase equilibria for sCO$_2$ and key components of LC bio-oils are very scarce. The fact that a large fraction of LC bio-oils is not fully characterized adds to this complexity.

The available solubility data of key components of LC bio-oils in the sCO$_2$-rich phase are correlated with the semi-empirical density-based model of Chrastil [137]. This procedure proved satisfactory for systems composed of sCO$_2$ and bio-oil components by Maqbool et al. [101] and, in this work, it is substantially extended to a larger number of components and chemical classes. The equation suggested by Chrastil takes the following form:

$$S = \rho^k \exp\left(\frac{a}{T} + b\right) \tag{1}$$

where:

S: solubility of the solute in the sCO$_2$-rich phase (grams of solute per liters of pure solvent at the P, T conditions of interest);
ρ: density of the pure solvent (g/L) at the P, T conditions of interest;
k: association number (i.e., number of CO$_2$ molecules associated with one solute molecule);
a: constant connected to the enthalpy of solvation and vaporization;
T: temperature (K);
b: constant connected to the molecular weights of the solvent and the solute.

The basic assumption of the Chrastil model is that a molecule of the solute is associated with a fixed number (k) of solvent molecules at a given temperature. The correlation proved successful for many components dissolved in sCO$_2$ corresponding to or having a similar structure to species found in LC bio-oils (e.g., phenol, naphthalene, octadecenoic acid [101,137]). In addition, it proved successful in a wide range of pressures and temperatures, provided that the solubility of the solute is not too high (typically solubility S not above 200 g/L) [137]. Considering that the VPLs of LC bio-oils, in the range of operating conditions of interest (see Section 3), are typically below 100 g/kg, the Chrastil model is expected to be a valid tool for correlation and data analysis.

Binary Phase Equilibrium Data of LC Bio-Oil Components and sCO$_2$

Table 9 reports a summary of the binary phase equilibrium data available in the literature, with reference to relevant components of LC bio-oils and sCO$_2$. For each publication, the component is specified, together with the pressure and temperature range of the measurements, as well as the corresponding density of pure sCO$_2$. The method of measurement is reported as analytical or synthetic, corresponding to the classification introduced by Dohrn and Brunner [138]. In short, the analytical method indicates that sampling and consequent analysis of the phases were performed, whereas in the synthetic method, bubble or dew point pressures are determined at given temperatures and overall compositions. With regard to the type of data, the symbol VLE indicates that both the solubility of the bio-oil components in the sCO$_2$-rich phase and the solubility of CO$_2$ in the liquid phase are reported. When "solubility" is indicated, the solubility of the bio-oil component in the sCO$_2$-rich phase is the only type of data reported. The complete solubility data can be found in the Supporting Information (Table S1). Table S1 collects all numerical data that are reported by the listed publications at supercritical sCO$_2$ conditions.

Table 9. Chemical components studied and experimental conditions of measured phase equilibrium data with sCO$_2$ in the literature. Temperature range (T), pressure range (P), sCO$_2$ density range (ρ), type of measurement (method), and type of data (data) are reported.

Component	T (°C)	P (bar)	ρ (kg/m^3)[1]	Method	Data	Ref.
Cyclohexanone	160–180	90–220	115–332	Analytical	VLE	[139]
5-Hydroxymethylfurfural	41–70	97–196	390–823	Synthetic	Solubility	[140]
Heptanoic acid	40–60	85–200	212–840	Analytical	Solubility	[141]
Hexadecanoic acid	40	80–248	278–878	Analytical	Solubility	[142]
	35–55	138–414	610–977	Analytical	Solubility	[143]
	35	99–230	709–888	Analytical	Solubility	[144]
	35–55	128–226	560–885	Analytical	Solubility	[145]
	40–45	101–233	512–864	Synthetic	Solubility	[146]
	64–78	105–260	643–759	Synthetic	VLE	[147]
Cumene	40–120	76–183	131–376	Analytical	VLE	[148]
	70	87–116	197–324	Analytical	VLE	[149]
	50	80–88	220–270	Analytical	VLE	[150]
Naphthalene	35–65	81–287	208–903	Analytical	Solubility	[151]
	121–162	77–166	120–313	Analytical	VLE	[152]
Benzeneacetic acid	35–45	90–200	381–852	Analytical	Solubility	[153]
Benzoic acid	35–70	101–364	252–958	Analytical	Solubility	[154]
	45–65	120–280	384–878	Analytical	Solubility	[155]
Phenol	60–90	100–350	203–863	Analytical	Solubility	[156]
	36–60	79–249	334–897	Analytical	Solubility	[157]
	100	107–301	207–663	Analytical	VLE	[158]
Catechol	60–90	100–350	203–863	Analytical	Solubility	[156]
	35–65	122–405	396–974	Analytical	Solubility	[159]
Vanillin	40–80	80–277	160–895	Analytical	Solubility	[160]
	35–45	83–195	466–857	Analytical	Solubility	[153]
	68–136	216–1341	561–1115	Synthetic	VLE	[161]
Guaiacol	50–120	80–200	128–784	Analytical	VLE	[162]
o-Cresol	100	104–263	199–610	Analytical	VLE	[158]
	50–200	99–300	121–848	Analytical	VLE	[163]
m-Cresol	35–55	80–240	204–895	Analytical	VLE	[164]
	100	102–300	194–662	Analytical	VLE	[158]
p-Cresol	50–200	100–348	123–898	Analytical	VLE	[163]
	80–150	80–200	113–594	Analytical	VLE	[162]
	100	103–302	196–663	Analytical	VLE	[158]

[1] Taken from NIST [109].

The components reported in Table 9 are chosen to represent the chemical classes indicated in Table 8, taking into account the availability of phase equilibrium data in the literature. All data retrieved in the literature were included in the analysis, with the exception of a few data sets concerning hexadecanoic acid and naphthalene. With regard to the former, six works were selected owing to the fact they show consistent data at the same P and T conditions, besides being based on high purity solutes and sound methodology. With regard to naphthalene, several literature sources exist. In this case, the work of McHugh and Paulaitis [151] was selected as their data are often used as reference by other works. In addition, the data reported by Yanagiuchi et al. [152] were chosen because they report data at high temperatures, as opposed to the typical range of 35–80 °C for all other published works.

Solubility data of LC bio-oil components in the sCO$_2$-rich phase were correlated using the linearized form of the Chrastil equation:

$$\ln(S) = k\ln(\rho) + \left(\frac{a}{T} + b\right) \tag{2}$$

Equation (2) was thus used for performing a multiple linear regression analysis of the experimental data reported in Table 9, referring to the sCO$_2$-rich phase. The density of pure CO_2 is taken from the NIST webbook, whose data are based on the Span–Wagner equation of state [109]. The set of optimal values for the parameters k, a, and b is reported for each component in Table 10, together with the regression statistics.

Table 10. Chrastil solubility constants (k, a, b) for the binary systems of bio-oil components and sCO$_2$, statistical properties of the regression (R^2, ARD), and number of experimental data regressed (N).

Component	Regressed Parameter			Goodness of Fit		
	k	a (K)	b	R^2	ARD %	N
Cyclohexanone	2.0216	−3425.3	0.45405	0.984	2.0	10
5-Hydroxymethylfurfural	4.0412	−3263.6	−15.267	0.973	9.0	19
Heptanoic acid	6.0527	−3806.1	−23.733	0.987	8.1	15
Hexadecanoic acid	7.5664	−9042.3	−20.706	0.950	19.0	62
Cumene	2.5852	−3481.6	−2.1835	0.963	9.3	41
Naphthalene	3.7852	−4080.9	−8.1975	0.940	14.7	63
Benzeneacetic acid	6.1072	−10177	−5.2314	0.986	4.9	24
Benzoic acid	5.2174	−5860.4	−14.636	0.986	13.4	53
Phenol	3.0544	−3081.4	6.9331	0.729	10.8	73
Catechol	3.7457	−3716.1	−12.417	0.977	18.2	62
Vanillin	4.0675	−4210.7	−11.707	0.948	18.4	74
Guaiacol	4.0447	−2597.8	−14.042	0.965	21.0	13
o-Cresol	3.4937	−3026.3	−9.2413	0.867	13.1	16
m-Cresol	3.8196	−2950.5	−12.383	0.98	18.0	23
p-Cresol	3.2734	−3441.3	−7.5747	0.924	15.7	30

For most data, the linearity is good, with R^2 values above 0.92. Exceptions are the values for phenol and o-cresol. For the latter, it is because of several solubility values being above 100–200 g/L. In this case, the density of the mixture starts departing from the solvent density and the model becomes less accurate [137]. In all cases, the average relative deviation between the calculated and experimental values is between 2% and 21%. The association value k is generally in line with available literature values [101,137,140,143]. It is observed that phenolics have values between 3 and 4, which is in line with the values reported by Maqbool et al. [101], even though the datasets are substantially expanded in this work. Cumene (1-ring aromatic hydrocarbon) exhibits one of the lowest k (i.e., 2.6), while naphthalene exhibits a higher value (i.e., 3.8), which is in line with its higher molecular size. Fatty acids have higher values, increasing with the molecular size as well.

Experimental solubilities of LC bio-oil components in the sCO$_2$-rich phase, expressed as grams of solute per kg of solvent, are plotted in Figure 9 as a function of the sCO$_2$ density for different temperatures. In the same figure, isotherms predicted by the Chrastil model are shown (continuous lines) for the corresponding values of temperature and solvent density. When more than one data source for the same experimental temperature exists, the one with the largest solvent density range was plotted.

In Figure 9, the effect of density and temperature can be appreciated. It is noted that increasing the solvent density at a given temperature corresponds to increasing pressures. As expected, the solubility increases with the solvent density for all isotherms. In addition, in all cases, higher temperatures at constant density conditions (i.e., increasing pressure) result in higher solubilities.

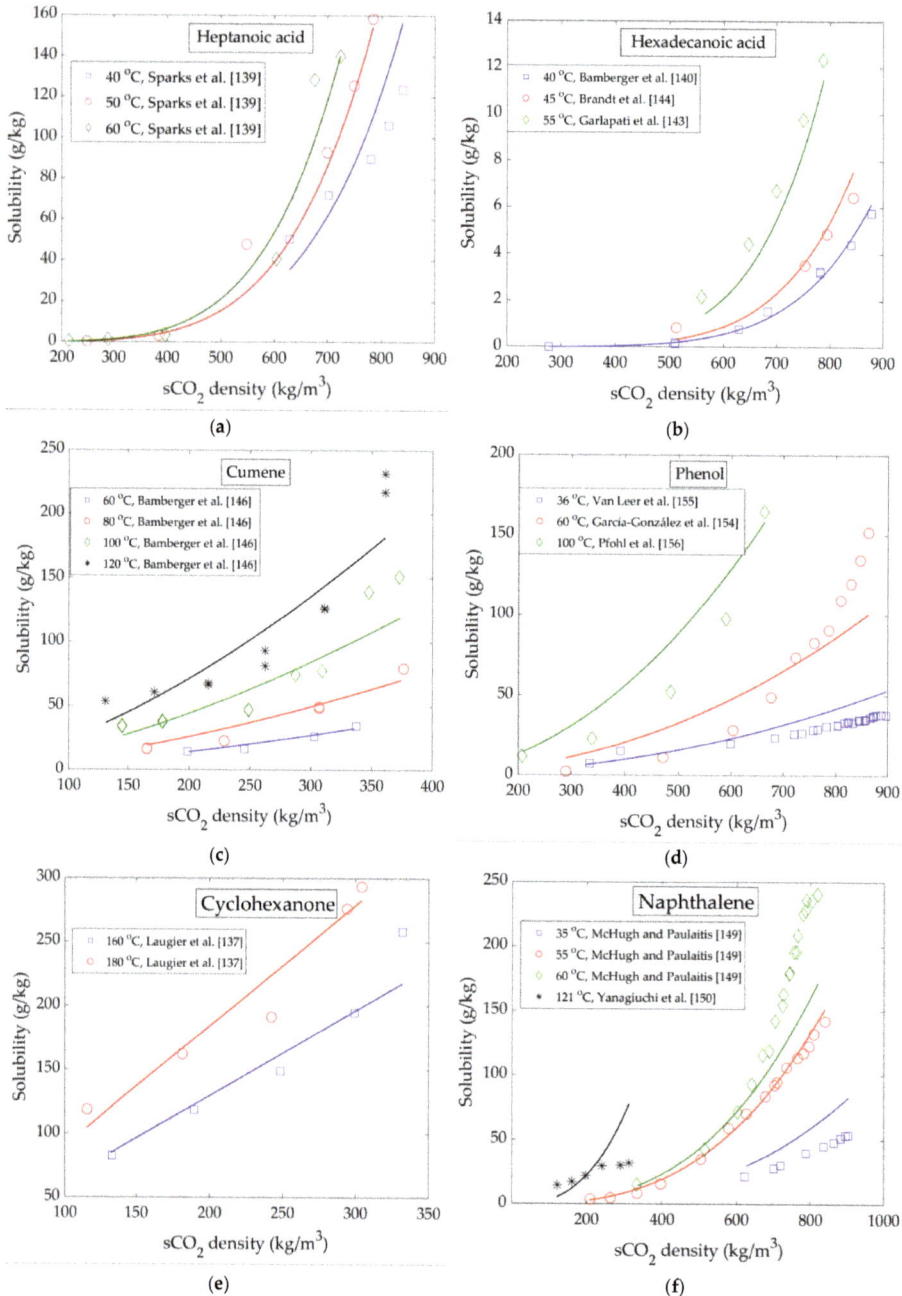

Figure 9. *Cont.*

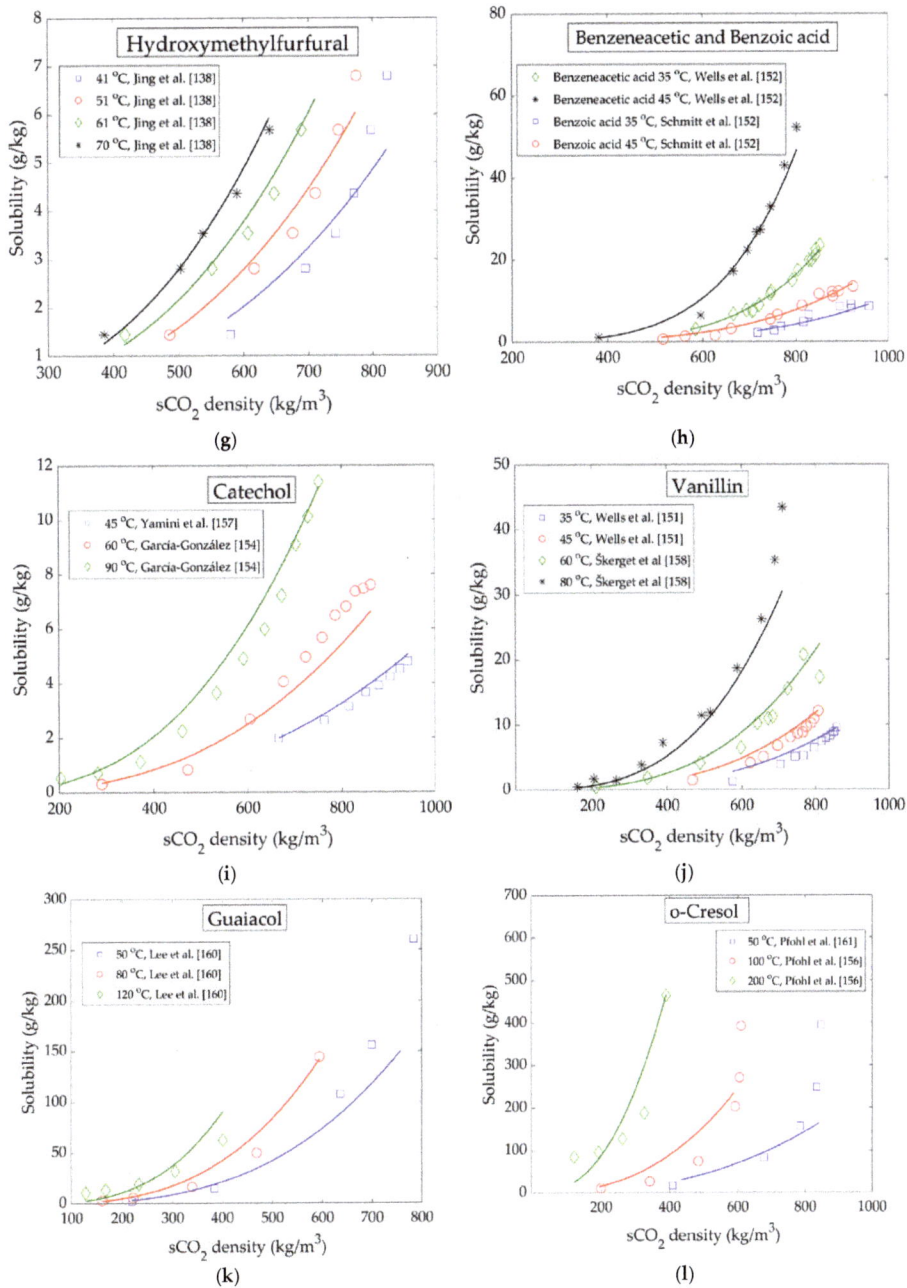

Figure 9. Cont.

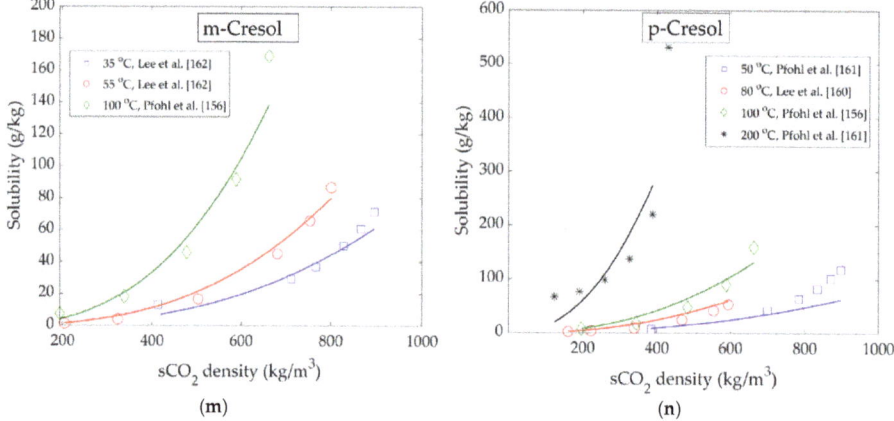

Figure 9. Solubility isotherms vs. sCO$_2$ density for all components studied. Markers indicate experimental data, while continuous lines are obtained with the Chrastil model. (**a**) Heptanoic acid; (**b**) Hexadecanoic acid; (**c**) Cumene; (**d**) Phenol; (**e**) Cyclohexanone; (**f**) Naphthalene; (**g**) Hydroxymethylfurfural; (**h**) Benzeneacetic and benzoic acid; (**i**) Catechol; (**j**) Vanillin; (**k**) Guaiacol; (**l**) o-Cresol; (**m**) m-Cresol; (**n**) p-Cresol.

These trends are in line with VPL values of sCO$_2$ extraction of LC bio-oils discussed in Section 3.3. A good example of the effect of temperature is vanillin, whose solubility at 80 °C increases considerably with the increasing solvent density, while at the lower temperatures, the increase is much less significant.

The solubility of sCO$_2$ in the liquid phase is provided only in a few cases. The available data show that sCO$_2$ exhibits high solubility in liquid hydrocarbons. For example, the solubility of sCO$_2$ in cumene ranges from 165 to 586 g/kg at 120 °C [148]. At the same temperature, naphthalene can dissolve 100–303 g/kg of CO$_2$ for pressure ranging from approximately 80 to 160 bar [152] CO$_2$. For the system sCO$_2$ + cyclohexanone, at the relatively high temperatures reported (i.e., 160 °C and 180 °C), CO$_2$ solubilities range from 146 to 555 g/kg [139]. A relatively large amount of CO$_2$ is dissolved in hexadecanoic acid as well, with the range being from 244 to 415 g/kg [147]. The solubility of CO$_2$ in phenolic components follows the trend cresols and guaiacol > phenol > vanillin [158–164]. In all cases, however, values above 380 g/kg were reported at the higher pressures (200–300 bar), with the exception of vanillin, where such high values were only reached at extremely high pressures (i.e., 762–1357 bar) [161]. In general, the data show that a large amount of sCO$_2$ can be dissolved in this type of liquid components. This is a favorable property for the separation of LC bio-oils in countercurrent equipment, as the dissolution of the supercritical solvent in the liquid causes the viscosity to drop, thus improving process operability.

5. Conclusions and Perspectives

The current work reviews the state-of-the-art on sCO$_2$ extraction of second-generation lignocellulosic bio-oils, produced by pyrolysis or hydrothermal liquefaction. sCO$_2$ extraction is deemed a promising physical separation process that can be integrated in the downstream of the thermochemical conversion unit with the aim of valorizing the raw bio-oils. Two main application areas are envisaged: (1) high yield extraction of HTL biocrudes for drop in bio-fuel production; (2) extraction of specific chemicals from pyrolysis oils.

In the case of HTL biocrudes, the application of sCO$_2$ allows extraction yields above 50%, with the production of extracts with favorable properties towards hydrotreatment. Namely, extracts have lower water content, lower average molecular weight, lower density, viscosity, and acidity, with the acidity nature shifted towards carboxylic instead of phenolic. In addition, the metal content of sCO$_2$ extracts is drastically reduced, from the very high values of HTL biocrudes (e.g., 8500 mg/kg) to values below 200 mg/kg, as metals are left in the residue of the sCO$_2$ extraction processes [58]. These properties are expected to lead to a lower deactivation rate of the hydrotreatment catalyst and reduced coking and fouling, overall leading to better process operability and economics. Moreover, the hydrotreatment products are expected to be shifted towards lighter hydrocarbon fractions, which are more valuable. On the other hand, the oxygen content of the extracts is comparable to that of the feed, or only moderately decreased, which means that significantly lower hydrogen requirements (per unit hydrotreatment feed) are not expected. However, this aspect is comparable to the experimental findings of vacuum distillation of HTL bio-oils, which also reported oxygen to be widespread in the whole range of distilled fractions. In this context, experimental works focusing on the comparison of hydrotreating of raw HTL biocrudes versus sCO$_2$ extracts is an important knowledge gap. Future research work is needed in this area.

In the case of pyrolysis oils, sCO$_2$ can be used for the recovery of fractions rich in short chain organic acids, such as acetic and propionic acids, short chain fatty acids, acetol, and furfural, owing to their relative abundance in this type of oil. The recovery of fractions rich in 1-ring lignin-derived phenolics (phenols and guaiacols) and their separation from benzenediols and long chain fatty acids is also deemed feasible on the basis of K-value analysis. High value chemicals like vanillin can also be recovered and made available for downstream purification. On the other hand, the sCO$_2$ residue of pyrolysis oil is concentrated in sugars and might have the potential to be fed to other biorefinery processes such as anaerobic digestion. In the case of pyrolysis oils, the use of sCO$_2$ for the downstream separation of the extracted components into several classes is a topic requiring further research in order to provide basic process schemes that can be evaluated in terms of process economics.

For both types of oils, the knowledge of the heavy fraction is scarce. Further studies addressing this aspect would be needed, in order to envisage possible utilizations of the residue, other than burning it for heat recovery. For example, it may have similarities with biochar and hydrochar, and can provide the base material for the production of adsorbents [165]. Phenolic oligomers from bio-oils were suggested as potential fuel antioxidants [105,107,108]. As sCO$_2$ extracts preferentially phenolic monomers, the residue is expected to be enriched in these components. The possibility of recovering this fraction using sCO$_2$, as pure solvent or with the use of small quantities of modifiers (e.g., ethanol, propanol), is another area of research to be developed.

As the natural competitor of sCO$_2$ extraction on LC bio-oils is vacuum distillation, experimental works aimed at comparing the two processes would be relevant to assess the advantages and drawbacks of the two processes in terms of yields and selectivity, as well as energy requirements for continuous countercurrent processing at a large scale.

Supplementary Materials: The following are available online at http://www.mdpi.com/1996-1073/13/7/1600/s1, Literature solubility of bio-oil components in sCO$_2$, Table S1.

Author Contributions: Conceptualization, N.M. and M.M.; methodology, N.M. and M.M.; writing—original draft preparation, N.M.; writing—review and editing, N.M. and M.M; Data Curation N.M.; Formal Analysis: N.M. and M.M.; Investigation: N.M. and M.M.; Visualization: N.M.; supervision, M.M. All authors have read and agreed to the published version of the manuscript.

Funding: This research received no external funding.

Conflicts of Interest: The authors declare no conflict of interest.

References

1. International Energy Outlook 2019. Available online: https://www.eia.gov/outlooks/ieo/ (accessed on 17 December 2019).
2. IEA. World Energy Statistics 2019. IEA: Paris, France, 2019. Available online: https://www.iea.org/reports/world-energy-statistics-2019 (accessed on 1 September 2019).

3. Kosinkova, J.; Doshi, A.; Maire, J.; Ristovski, Z.; Brown, R.; Rainey, T.J. Measuring the regional availability of biomass for biofuels and the potential for microalgae. *Renew. Sustain. Energy Rev.* **2015**, *49*, 1271–1285. [CrossRef]
4. US Environmental Protection Agency. *Biofuels and the Environment: Second Triennial Report to Congress*; US Environmental Protection Agency: Washington, DC, USA, 2018.
5. IEA Renewables 2019. Available online: https://www.iea.org/reports/renewables-2019 (accessed on 17 December 2019).
6. Baloch, H.A.; Nizamuddin, S.; Siddiqui, M.T.H.; Riaz, S.; Jatoi, A.S.; Dumbre, D.K.; Mubarak, N.M.; Srinivasan, M.P.; Griffin, G.J. Recent advances in production and upgrading of bio-oil from biomass: A critical overview. *J. Environ. Chem. Eng.* **2018**, *6*, 5101–5118. [CrossRef]
7. Ramirez, J.A.; Brown, R.J.; Rainey, T.J. A review of hydrothermal liquefaction bio-crude properties and prospects for upgrading to transportation fuels. *Energies* **2015**, *8*, 6765–6794. [CrossRef]
8. Bridgwater, A.V. Review of fast pyrolysis of biomass and product upgrading. *Biomass Bioenergy* **2012**, *38*, 68–94. [CrossRef]
9. Hoffmann, J.; Jensen, C.U.; Rosendahl, L.A. Co-processing potential of HTL bio-crude at petroleum refineries-Part 1: Fractional distillation and characterization. *Fuel* **2016**, *165*, 526–535. [CrossRef]
10. Pedersen, T.H.; Jensen, C.U.; Sandström, L.; Rosendahl, L.A. Full characterization of compounds obtained from fractional distillation and upgrading of a HTL biocrude. *Appl. Energy* **2017**, *202*, 408–419. [CrossRef]
11. Elkasabi, Y.; Mullen, C.A.; Boateng, A.A. Distillation and isolation of commodity chemicals from bio-oil made by tail-gas reactive pyrolysis. *ACS Sustain. Chem. Eng.* **2014**, *2*, 2042–2052. [CrossRef]
12. Capunitan, J.A.; Capareda, S.C. Characterization and separation of corn stover bio-oil by fractional distillation. *Fuel* **2013**, *112*, 60–73. [CrossRef]
13. Taghipour, A.; Ramirez, J.A.; Brown, R.J.; Rainey, T.J. A review of fractional distillation to improve hydrothermal liquefaction biocrude characteristics; future outlook and prospects. *Renew. Sustain. Energy Rev.* **2019**, *115*, 109355. [CrossRef]
14. Hu, H.S.; Wu, Y.L.; Yang, M.D. Fractionation of bio-oil produced from hydrothermal liquefaction of microalgae by liquid-liquid extraction. *Biomass Bioenergy* **2018**, *108*, 487–500. [CrossRef]
15. Bjelić, S.; Yu, J.; Iversen, B.B.; Glasius, M.; Biller, P. Detailed investigation into the asphaltene fraction of hydrothermal liquefaction derived bio-crude and hydrotreated bio-crudes. *Energy and Fuels* **2018**, *32*, 3579–3587. [CrossRef]
16. Brunner, G. Counter-current separations. *J. Supercrit. Fluids* **2009**, *47*, 574–582. [CrossRef]
17. Nielsen, R.P.; Valsecchi, R.; Strandgaard, M.; Maschietti, M. Experimental study on fluid phase equilibria of hydroxyl-terminated perfluoropolyether oligomers and supercritical carbon dioxide. *J. Supercrit. Fluids* **2015**, *101*, 124–130. [CrossRef]
18. Speight, J.G. Nonthermal methods of recovery. In *Enhanced Recovery Methods for Heavy Oil and Tar Sands*; Elsevier: Austin, TX, USA, 2009; pp. 185–220. ISBN 9780127999883.
19. Kummamuru, B. *Global Bioenergy Statistics*; World Bioenergy Association: Stockholm, Sweden, 2017.
20. Mateo-Sagasta, J.; Raschid-Sally, L.; Thebo, A. Global Wastewater and Sludge Production, Treatment and Use. In *Wastewater: Economic Asset in an Urbanizing World*; Drechsel, P., Qadir, M., Wichelns, D., Eds.; Springer: Dordrecht, The Netherlands, 2015; pp. 15–38. ISBN 978-94-017-9545-6.
21. Lignin Products Global Market Size, Sales Data 2017-2022 & Applications in Animal Feed Industry. Available online: https://www.orbisresearch.com/contacts/request-sample/218258 (accessed on 13 May 2019).
22. Tian, X.; Fang, Z.; Smith, R.L.; Wu, Z.; Liu, M. Properties, chemical characteristics and application of lignin and its derivatives. In *Production of Biofuels and Chemicals from Lignin*; Fang, Z., Smith, R.L., Jr., Eds.; Springer: Singapore, 2016; pp. 3–33. ISBN 978-981-10-1965-4.
23. Zakzeski, J.; Bruijnincx, P.C.A.; Jongerius, A.L.; Weckhuysen, B.M. The catalytic valorization of lignin for the production of renewable chemicals. *Chem. Rev.* **2010**, *110*, 3552–3599. [CrossRef]
24. Miliotti, E.; Dell'Orco, S.; Lotti, G.; Rizzo, A.M.; Rosi, L.; Chiaramonti, D. Lignocellulosic ethanol biorefinery: Valorization of lignin-rich stream through hydrothermal liquefaction. *Energies* **2019**, *12*, 723. [CrossRef]
25. Sumathi, S.; Chai, S.P.; Mohamed, A.R. Utilization of oil palm as a source of renewable energy in Malaysia. *Renew. Sustain. Energy Rev.* **2008**, *12*, 2404–2421. [CrossRef]
26. ECN.TNO Phyllis2, Database for Biomass and Waste. Available online: https://phyllis.nl/ (accessed on 18 November 2019).

27. Oasmaa, A.; Van De Beld, B.; Saari, P.; Elliott, D.C.; Solantausta, Y. Norms, standards, and legislation for fast pyrolysis bio-oils from lignocellulosic biomass. *Energy Fuels* **2015**, *29*, 2471–2484. [CrossRef]
28. Nguyen, T.D.H.; Maschietti, M.; Åmand, L.E.; Vamling, L.; Olausson, L.; Andersson, S.I.; Theliander, H. The effect of temperature on the catalytic conversion of Kraft lignin using near-critical water. *Bioresour. Technol.* **2014**, *170*, 196–203. [CrossRef]
29. Feng, S.; Yuan, Z.; Leitch, M.; Xu, C.C. Hydrothermal liquefaction of barks into bio-crude - Effects of species and ash content/composition. *Fuel* **2014**, *116*, 214–220. [CrossRef]
30. Haarlemmer, G.; Guizani, C.; Anouti, S.; Déniel, M.; Roubaud, A.; Valin, S. Analysis and comparison of bio-oils obtained by hydrothermal liquefaction and fast pyrolysis of beech wood. *Fuel* **2016**, *174*, 180–188. [CrossRef]
31. Chen, Y.; Cao, X.; Zhu, S.; Tian, F.; Xu, Y.; Zhu, C.; Dong, L. Synergistic hydrothermal liquefaction of wheat stalk with homogeneous and heterogeneous catalyst at low temperature. *Bioresour. Technol.* **2019**, *278*, 92–98. [CrossRef]
32. Baloch, H.A.; Nizamuddin, S.; Siddiqui, M.T.H.; Mubarak, N.M.; Dumbre, D.K.; Srinivasan, M.P.; Griffin, G.J. Sub-supercritical liquefaction of sugarcane bagasse for production of bio-oil and char: Effect of two solvents. *J. Environ. Chem. Eng.* **2018**, *6*, 6589–6601. [CrossRef]
33. Castello, D.; Pedersen, T.; Rosendahl, L. Continuous hydrothermal liquefaction of biomass: A critical review. *Energies* **2018**, *11*, 3165. [CrossRef]
34. BTG-BTL Empyro Project. Available online: https://www.btg-btl.com/en/company/projects/empyro (accessed on 22 October 2019).
35. Ensyn Côte-Nord. Available online: http://www.ensyn.com/quebec.html (accessed on 22 October 2019).
36. Silva Green Fuel. Available online: https://www.statkraft.com/about-statkraft/Projects/norway/value-creation-tofte/silva-green-fuel/ (accessed on 22 October 2019).
37. Elliott, D.C.; Oasmaa, A.; Meier, D.; Preto, F.; Bridgwater, A.V. Results of the IEA round robin on viscosity and aging of fast pyrolysis bio-oils: Long-Term tests and repeatability. *Energy Fuels* **2012**, *26*, 7362–7366. [CrossRef]
38. Ong, B.H.Y.; Walmsley, T.G.; Atkins, M.J.; Walmsley, M.R.W. Hydrothermal liquefaction of Radiata Pine with Kraft black liquor for integrated biofuel production. *J. Clean. Prod.* **2018**, *199*, 737–750. [CrossRef]
39. Jarvis, J.M.; Albrecht, K.O.; Billing, J.M.; Schmidt, A.J.; Hallen, R.T.; Schaub, T.M. Assessment of hydrotreatment for hydrothermal liquefaction biocrudes from sewage sludge, microalgae, and pine feedstocks. *Energy Fuels* **2018**, *32*, 8483–8493. [CrossRef]
40. Belkheiri, T.; Andersson, S.I.; Mattsson, C.; Olausson, L.; Theliander, H.; Vamling, L. Hydrothermal liquefaction of kraft lignin in sub-critical water: the influence of the sodium and potassium fraction. *Biomass Convers. Biorefinery* **2018**, *8*, 585–595. [CrossRef]
41. Anastasakis, K.; Biller, P.; Madsen, R.B.; Glasius, M.; Johannsen, I. Continuous hydrothermal liquefaction of biomass in a novel pilot plant with heat recovery and hydraulic oscillation. *Energies* **2018**, *11*, 2695. [CrossRef]
42. Gollakota, A.R.K.; Kishore, N.; Gu, S. A review on hydrothermal liquefaction of biomass. *Renew. Sustain. Energy Rev.* **2018**, *81*, 1378–1392. [CrossRef]
43. Feng, Y.; Meier, D. Supercritical carbon dioxide extraction of fast pyrolysis oil from softwood. *J. Supercrit. Fluids* **2017**, *128*, 6–17. [CrossRef]
44. Leijenhorst, E.J.; Wolters, W.; Van De Beld, L.; Prins, W. Inorganic element transfer from biomass to fast pyrolysis oil: Review and experiments. *Fuel Process. Technol.* **2016**, *149*, 96–111. [CrossRef]
45. Sintamarean, I.M.; Grigoras, I.F.; Jensen, C.U.; Toor, S.S.; Pedersen, T.H.; Rosendahl, L.A. Two-stage alkaline hydrothermal liquefaction of wood to biocrude in a continuous bench-scale system. *Biomass Convers. Biorefinery* **2017**, *7*, 425–435. [CrossRef]
46. Kosinkova, J.; Ramirez, J.A.; Ristovski, Z.D.; Brown, R.; Rainey, T.J. Physical and chemical stability of bagasse biocrude from liquefaction stored in real conditions. *Energy Fuels* **2016**, *30*, 10499–10504. [CrossRef]
47. Elliott, D.C.; Meier, D.; Oasmaa, A.; Van De Beld, B.; Bridgwater, A.V.; Marklund, M. Results of the international energy agency round robin on fast pyrolysis bio-oil production. *Energy Fuels* **2017**, *31*, 5111–5119. [CrossRef]
48. Speight, J.G. Petroleum Analysis. In *The Chemistry and Technology of Petroleum*; CRC Press: Boca Raton, FL, USA, 2006; pp. 274–314. ISBN 9780429118494.

49. Fahmi, R.; Bridgwater, A.V.; Donnison, I.; Yates, N.; Jones, J.M. The effect of lignin and inorganic species in biomass on pyrolysis oil yields, quality and stability. *Fuel* **2008**, *87*, 1230–1240. [CrossRef]
50. Das, P.; Ganesh, A.; Wangikar, P. Influence of pretreatment for deashing of sugarcane bagasse on pyrolysis products. *Biomass Bioenergy* **2004**, *27*, 445–457. [CrossRef]
51. Minowa, T.; Kondo, T.; Sudirjo, S.T. Thermochemical liquefaction of Indonesian biomass residues. *Biomass Bioenergy* **1998**, *14*, 517–524. [CrossRef]
52. Déniel, M.; Haarlemmer, G.; Roubaud, A.; Weiss-Hortala, E.; Fages, J. Optimisation of bio-oil production by hydrothermal liquefaction of agro-industrial residues: Blackcurrant pomace (Ribes nigrum L.) as an example. *Biomass Bioenergy* **2016**, *95*, 273–285. [CrossRef]
53. Speight, J.G. Chemical Composition. In *The Chemistry and Technology of Petroleum*; Chemical Industries; CRC Press: Boca Raton, FL, USA, 2014; ISBN 9781439873892.
54. Ghorbannezhad, P.; Kool, F.; Rudi, H.; Ceylan, S. Sustainable production of value-added products from fast pyrolysis of palm shell residue in tandem micro-reactor and pilot plant. *Renew. Energy* **2020**, *145*, 663–670. [CrossRef]
55. Gómez-Monedero, B.; Bimbela, F.; Arauzo, J.; Faria, J.; Ruiz, M.P. Pyrolysis of red eucalyptus, camelina straw, and wheat straw in an ablative reactor. *Energy Fuels* **2015**, *29*, 1766–1775. [CrossRef]
56. Hernando, H.; Jiménez-Sánchez, S.; Fermoso, J.; Pizarro, P.; Coronado, J.M.; Serrano, D.P. Assessing biomass catalytic pyrolysis in terms of deoxygenation pathways and energy yields for the efficient production of advanced biofuels. *Catal. Sci. Technol.* **2016**, *6*, 2829–2843. [CrossRef]
57. Pedersen, T.H.; Grigoras, I.F.; Hoffmann, J.; Toor, S.S.; Daraban, I.M.; Jensen, C.U.; Iversen, S.B.; Madsen, R.B.; Glasius, M.; Arturi, K.R.; et al. Continuous hydrothermal co-liquefaction of aspen wood and glycerol with water phase recirculation. *Appl. Energy* **2016**, *162*, 1034–1041. [CrossRef]
58. Montesantos, N.; Nielsen, R.P.; Maschietti, M. Upgrading of nondewatered nondemetallized lignocellulosic biocrude from hydrothermal liquefaction using supercritical carbon dioxide. *Ind. Eng. Chem. Res.* **2020**, accepted. [CrossRef]
59. Elliott, D.C.; Oasmaa, A.; Preto, F.; Meier, D.; Bridgwater, A.V. Results of the IEA round robin on viscosity and stability of fast pyrolysis bio-oils. *Energy Fuels* **2012**, *26*, 3769–3776. [CrossRef]
60. Jensen, C.U.; Rodriguez Guerrero, J.K.; Karatzos, S.; Olofsson, G.; Iversen, S.B. Fundamentals of Hydrofaction™: Renewable crude oil from woody biomass. *Biomass Convers. Biorefinery* **2017**, *7*, 495–509. [CrossRef]
61. Jarvis, J.M.; Billing, J.M.; Hallen, R.T.; Schmidt, A.J.; Schaub, T.M. Hydrothermal liquefaction biocrude compositions compared to petroleum crude and shale oil. *Energy Fuels* **2017**, *31*, 2896–2906. [CrossRef]
62. Madsen, R.B.; Bernberg, R.Z.K.; Biller, P.; Becker, J.; Iversen, B.B.; Glasius, M. Hydrothermal co-liquefaction of biomasses-quantitative analysis of bio-crude and aqueous phase composition. *Sustain. Energy Fuels* **2017**, *1*, 789–805. [CrossRef]
63. Christensen, E.D.; Chupka, G.M.; Luecke, J.; Smurthwaite, T.; Alleman, T.L.; Iisa, K.; Franz, J.A.; Elliott, D.C.; McCormick, R.L. Analysis of oxygenated compounds in hydrotreated biomass fast pyrolysis oil distillate fractions. *Energy Fuels* **2011**, *25*, 5462–5471. [CrossRef]
64. Montesantos, N.; Pedersen, T.H.; Nielsen, R.P.; Rosendahl, L.; Maschietti, M. Supercritical carbon dioxide fractionation of bio-crude produced by hydrothermal liquefaction of pinewood. *J. Supercrit. Fluids* **2019**, *149*, 97–109. [CrossRef]
65. ASTM International. *ASTM D664-17a. Standard Test Method for Acid Number of Petroleum Products by Potentiometric Titration*; ASTM: West Conshohocken, PA, USA, 2017.
66. Jensen, C.U.; Rosendahl, L.A.; Olofsson, G. Impact of nitrogenous alkaline agent on continuous HTL of lignocellulosic biomass and biocrude upgrading. *Fuel Process. Technol.* **2017**, *159*, 376–385. [CrossRef]
67. Jacobson, K.; Maheria, K.C.; Kumar Dalai, A. Bio-oil valorization: A review. *Renew. Sustain. Energy Rev.* **2013**, *23*, 91–106. [CrossRef]
68. Stankovikj, F.; McDonald, A.G.; Helms, G.L.; Garcia-Perez, M. Quantification of bio-oil functional groups and evidences of the presence of pyrolytic humins. *Energy Fuels* **2016**, *30*, 6505–6524. [CrossRef]
69. Madsen, R.B.; Anastasakis, K.; Biller, P.; Glasius, M. Rapid determination of water, total acid number, and phenolic content in bio-crude from hydrothermal liquefaction of biomass using FT-IR. *Energy Fuels* **2018**, *32*, 7660–7669. [CrossRef]

70. Harman-Ware, A.E.; Ferrell, J.R. Methods and challenges in the determination of molecular weight metrics of bio-oils. *Energy Fuels* **2018**, *32*, 8905–8920. [CrossRef]
71. Hwang, H.; Lee, J.H.; Choi, I.G.; Choi, J.W. Comprehensive characterization of hydrothermal liquefaction products obtained from woody biomass under various alkali catalyst concentrations. *Environ. Technol. (United Kingdom)* **2019**, *40*, 1657–1667. [CrossRef]
72. Belkheiri, T.; Vamling, L.; Nguyen, T.D.H.; Maschietti, M.; Olausson, L.; Andersson, S.-I.; Åmand, L.-E.; Theliander, H. Kraft lignin depolymerization in near-critical water: effect of changing co-solvent. *Cell Chem. Technol.* **2014**, *48*, 813–818.
73. Lyckeskog, H.N.; Mattsson, C.; Olausson, L.; Andersson, S.I.; Vamling, L.; Theliander, H. Thermal stability of low and high Mw fractions of bio-oil derived from lignin conversion in subcritical water. *Biomass Convers. Biorefinery* **2017**, *7*, 401–414. [CrossRef]
74. Conrad, S.; Blajin, C.; Schulzke, T.; Deerberg, G. Comparison of fast pyrolysis bio-oils from straw and miscanthus. *Environ. Prog. Sustain. Energy* **2019**, *38*, e13287. [CrossRef]
75. Yildiz, G.; Ronsse, F.; Venderbosch, R.; van Duren, R.; Kersten, S.R.A.; Prins, W. Effect of biomass ash in catalytic fast pyrolysis of pine wood. *Appl. Catal. B Environ.* **2015**, *168–169*, 203–211. [CrossRef]
76. Doassans-Carrère, N.; Ferrasse, J.H.; Boutin, O.; Mauviel, G.; Lédé, J. Comparative study of biomass fast pyrolysis and direct liquefaction for bio-oils production: Products yield and characterizations. *Energy Fuels* **2014**, *28*, 5103–5111. [CrossRef]
77. Azargohar, R.; Jacobson, K.L.; Powell, E.E.; Dalai, A.K. Evaluation of properties of fast pyrolysis products obtained, from Canadian waste biomass. *J. Anal. Appl. Pyrolysis* **2013**, *104*, 330–340. [CrossRef]
78. Undri, A.; Abou-Zaid, M.; Briens, C.; Berruti, F.; Rosi, L.; Bartoli, M.; Frediani, M.; Frediani, P. A simple procedure for chromatographic analysis of bio-oils from pyrolysis. *J. Anal. Appl. Pyrolysis* **2015**, *114*, 208–221. [CrossRef]
79. Nguyen Lyckeskog, H.; Mattsson, C.; Åmand, L.E.; Olausson, L.; Andersson, S.I.; Vamling, L.; Theliander, H. Storage stability of bio-oils derived from the catalytic conversion of softwood Kraft lignin in subcritical water. *Energy Fuels* **2016**, *30*, 3097–3106. [CrossRef]
80. Nguyen, T.D.H.; Maschietti, M.; Belkheiri, T.; Åmand, L.E.; Theliander, H.; Vamling, L.; Olausson, L.; Andersson, S.I. Catalytic depolymerisation and conversion of Kraft lignin into liquid products using near-critical water. *J. Supercrit. Fluids* **2014**, *86*, 67–75. [CrossRef]
81. Maddi, B.; Viamajala, S.; Varanasi, S. Comparative study of pyrolysis of algal biomass from natural lake blooms with lignocellulosic biomass. *Bioresour. Technol.* **2011**, *102*, 11018–11026. [CrossRef]
82. Aqsha, A.; Tijani, M.M.; Moghtaderi, B.; Mahinpey, N. Catalytic pyrolysis of straw biomasses (wheat, flax, oat and barley) and the comparison of their product yields. *J. Anal. Appl. Pyrolysis* **2017**, *125*, 201–208. [CrossRef]
83. Zhu, Z.; Rosendahl, L.; Toor, S.S.; Yu, D.; Chen, G. Hydrothermal liquefaction of barley straw to bio-crude oil: Effects of reaction temperature and aqueous phase recirculation. *Appl. Energy* **2015**, *137*, 183–192. [CrossRef]
84. Feng, Y.; Meier, D. Extraction of value-added chemicals from pyrolysis liquids with supercritical carbon dioxide. *J. Anal. Appl. Pyrolysis* **2015**, *113*, 174–185. [CrossRef]
85. Li, C.; Zhao, X.; Wang, A.; Huber, G.W.; Zhang, T. Catalytic transformation of lignin for the production of chemicals and fuels. *Chem. Rev.* **2015**, *115*, 11559–11624. [CrossRef]
86. Nanda, S.; Mohammad, J.; Reddy, S.N.; Kozinski, J.A.; Dalai, A.K. Pathways of lignocellulosic biomass conversion to renewable fuels. *Biomass Convers. Biorefinery* **2014**, *4*, 157–191. [CrossRef]
87. Chan, Y.H.; Quitain, A.T.; Yusup, S.; Uemura, Y.; Sasaki, M.; Kida, T. Liquefaction of palm kernel shell in sub- and supercritical water for bio-oil production. *J. Energy Inst.* **2018**, *91*, 721–732. [CrossRef]
88. Kokayeff, P.; Zink, S.; Roxas, P. Hydrotreating in petroleum processing. In *Handbook of Petroleum Processing*; Treese, S.A., Pujadó, P.R., Jones, D.S.J., Eds.; Springer International Publishing: Berlin/Heidelberg, Germany, 1995; pp. 363–434. ISBN 978-3-319-14529-7.
89. Castello, D.; Haider, M.S.; Rosendahl, L.A. Catalytic upgrading of hydrothermal liquefaction biocrudes: Different challenges for different feedstocks. *Renew. Energy* **2019**, *141*, 420–430. [CrossRef]
90. Jensen, C.U.; Hoffmann, J.; Rosendahl, L.A. Co-processing potential of HTL bio-crude at petroleum refineries. Part 2: A parametric hydrotreating study. *Fuel* **2016**, *165*, 536–543. [CrossRef]
91. Gollakota, A.R.K.; Reddy, M.; Subramanyam, M.D.; Kishore, N. A review on the upgradation techniques of pyrolysis oil. *Renew. Sustain. Energy Rev.* **2016**, *58*, 1543–1568. [CrossRef]

92. Hoffmann, J.; Pedersen, T.H.; Rosendahl, L.A. Near-critical and supercritical water and their applications for biorefineries. In *Biofuels and Biorefineries*; Fang, Z., Xu, C., Eds.; Biofuels and Biorefineries; Springer: Dordrecht, The Netherlands, 2014; Volume 2, pp. 373–400. ISBN 978-94-017-8922-6.
93. Han, Y.; Gholizadeh, M.; Tran, C.-C.; Kaliaguine, S.; Li, C.-Z.; Olarte, M.; Garcia-Perez, M. Hydrotreatment of pyrolysis bio-oil: A review. *Fuel Process. Technol.* **2019**, *195*, 106140. [CrossRef]
94. Furimsky, E.; Massoth, F.E. Deactivation of hydroprocessing catalysts. *Catal. Today* **1999**, *52*, 381–495. [CrossRef]
95. Mohamad, M.H.; Awang, R.; Yunus, W.Z.W. A Review of acetol: Application and production. *Am. J. Appl. Sci.* **2011**, *8*, 1135–1139. [CrossRef]
96. Chen, L.; Zhao, J.; Pradhan, S.; Brinson, B.E.; Scuseria, G.E.; Zhang, Z.C.; Wong, M.S. Ring-locking enables selective anhydrosugar synthesis from carbohydrate pyrolysis. *Green Chem.* **2016**, *18*, 5438–5447. [CrossRef]
97. Rover, M.R.; Aui, A.; Wright, M.M.; Smith, R.G.; Brown, R.C. Production and purification of crystallized levoglucosan from pyrolysis of lignocellulosic biomass. *Green Chem.* **2019**, *21*, 5980–5989. [CrossRef]
98. Werpy, T.; Petersen, G. *Top Value Added Chemicals from Biomass: Volume I-Results of Screening for Potential Candidates from Sugars and Synthesis Gas*; National Renewable Energy Laboratory (NREL): Golden, CO, USA, 2004.
99. Longley, C.J.; Fung, D.P.C. Potential Applications and Markets for Biomass-Derived Levoglucosan. In *Advances in Thermochemical Biomass Conversion*; Springer: Dordrecht, The Netherlands, 1993; pp. 1484–1494. ISBN 978-94-011-1336-6.
100. What's New in Phenol Production? Available online: https://www.acs.org/content/acs/en/pressroom/cutting-edge-chemistry/what-s-new-in-phenol-production-.html (accessed on 12 August 2019).
101. Maqbool, W.; Hobson, P.; Dunn, K.; Doherty, W. Supercritical carbon dioxide separation of carboxylic acids and phenolics from bio-oil of lignocellulosic origin: Understanding bio-oil compositions, compound solubilities, and their fractionation. *Ind. Eng. Chem. Res.* **2017**, *56*, 3129–3144. [CrossRef]
102. Fiege, H.; Voges, H.-W.; Hamamoto, T.; Umemura, S.; Iwata, T.; Miki, H.; Fujita, Y.; Buysch, H.-J.; Garbe, D.; Paulus, W. Phenol Derivatives. In *Ullmann's Encyclopedia of Industrial Chemistry*; Wiley-VCH Verlag GmbH & Co. KGaA: Weinheim, Germany, 2012; Volume 26, pp. 552–553.
103. Naik, S.; Goud, V.V.; Rout, P.K.; Dalai, A.K. Supercritical CO_2 fractionation of bio-oil produced from wheat-hemlock biomass. *Bioresour. Technol.* **2010**, *101*, 7605–7613. [CrossRef] [PubMed]
104. Vanilla and Vanillin Market: Global Industry Trends, Share, Size, Growth, Opportunity and Forecast 2019-2024. Available online: https://www.researchandmarkets.com/research/n4fxw5/global_vanilla?w=12 (accessed on 28 October 2019).
105. Wu, X.F.; Zhou, Q.; Li, M.F.; Li, S.X.; Bian, J.; Peng, F. Conversion of poplar into bio-oil via subcritical hydrothermal liquefaction: Structure and antioxidant capacity. *Bioresour. Technol.* **2018**, *270*, 216–222. [CrossRef] [PubMed]
106. Larson, R.A.; Sharma, B.K.; Marley, K.A.; Kunwar, B.; Murali, D.; Scott, J. Potential antioxidants for biodiesel from a softwood lignin pyrolyzate. *Ind. Crops Prod.* **2017**, *109*, 476–482. [CrossRef]
107. Chandrasekaran, S.R.; Murali, D.; Marley, K.A.; Larson, R.A.; Doll, K.M.; Moser, B.R.; Scott, J.; Sharma, B.K. Antioxidants from slow pyrolysis bio-oil of birch wood: Application for biodiesel and biobased lubricants. *ACS Sustain. Chem. Eng.* **2016**, *4*, 1414–1421. [CrossRef]
108. Qazi, S.S.; Li, D.; Briens, C.; Berruti, F.; Abou-Zaid, M.M. Antioxidant activity of the lignins derived from fluidized-bed fast pyrolysis. *Molecules* **2017**, *22*, 372. [CrossRef] [PubMed]
109. Lemmon, E.W.; McLinden, M.O.; Friend, D.G. Thermophysical Properties of Fluid Systems. In *NIST Chemistry WebBook, NIST Standard Reference Database Number 69*; Linstrom, P.J., Mallard, W.G., Eds.; National Institute of Standards and Technology: Gaithersburg, MD, USA, 1998.
110. Michels, A.; Blaisse, B.; Hoogschagen, J. The melting line of carbon dioxide up to 2800 atmospheres. *Physica* **1942**, *9*, 565–573. [CrossRef]
111. Dortmund Data Bank. Available online: http://www.ddbst.com/en/EED/PCP/VAP_C1050.php (accessed on 29 October 2019).
112. Brunner, G. Supercritical fluids: Technology and application to food processing. *J. Food Eng.* **2005**, *67*, 21–33. [CrossRef]
113. McHugh, M.A.; Krukonis, V.J. *Supercritical Fluid Extraction*, 2nd ed.; Brenner, H., Ed.; Butterworth-Heinemann series in chemical engineering; Elsevier: Amsterdam, The Netherlands, 1994; ISBN 9780080518176.

114. Gupta, R.B.; Shim, J.-J. *Solubility in Supercritical Carbon Dioxide*; Gupta, R.B., Shim, J.-J., Eds.; CRC Press: Boca Raton, FL, USA, 2006; ISBN 9780429122088.
115. Reverchon, E.; De Marco, I. Supercritical fluid extraction and fractionation of natural matter. *J. Supercrit. Fluids* **2006**, *38*, 146–166. [CrossRef]
116. McHugh, M.A.; Krukonis, V.J. Processing Pharmaceuticals, Natural Products, Specialty Chemicals, and Waste Streams. In *Supercritical Fluid Extraction*; Butterworth-Heinemann: Oxford, UK, 1994; pp. 293–310. ISBN 9780080518176.
117. Gironi, F.; Maschietti, M. Supercritical carbon dioxide fractionation of lemon oil by means of a batch process with an external reflux. *J. Supercrit. Fluids* **2005**, *35*, 227–234. [CrossRef]
118. Maschietti, M.; Pedacchia, A. Supercritical carbon dioxide separation of fish oil ethyl esters by means of a continuous countercurrent process with an internal reflux. *J. Supercrit. Fluids* **2014**, *86*, 76–84. [CrossRef]
119. Gironi, F.; Maschietti, M. Continuous countercurrent deterpenation of lemon essential oil by means of supercritical carbon dioxide: Experimental data and process modelling. *Chem. Eng. Sci.* **2008**, *63*, 651–661. [CrossRef]
120. Gironi, F.; Maschietti, M. Separation of fish oils ethyl esters by means of supercritical carbon dioxide: Thermodynamic analysis and process modelling. *Chem. Eng. Sci.* **2006**, *61*, 5114–5126. [CrossRef]
121. Riha, V.; Brunner, G. Separation of fish oil ethyl esters with supercritical carbon dioxide. *J. Supercrit. Fluids* **2000**, *17*, 55–64. [CrossRef]
122. Osséo, L.S.; Caputo, G.; Gracia, I.; Reverchon, E. Continuous fractionation of used frying oil by supercritical CO_2. *JAOCS J. Am. Oil Chem. Soc.* **2004**, *81*, 879–885. [CrossRef]
123. Kim, S.K.; Han, J.Y.; Hong, S.A.; Lee, Y.W.; Kim, J. Supercritical CO_2-purification of waste cooking oil for high-yield diesel-like hydrocarbons via catalytic hydrodeoxygenation. *Fuel* **2013**, *111*, 510–518. [CrossRef]
124. Meyer, T. Extracting and upgrading heavy hydrocarbons using supercritical carbon dioxide 2011. UK Patent GB 2471862 A, 19 January 2011.
125. Subramanian, A.; Floyd, R. Residuum oil supercritical extraction process 2011. US Patent 2011/0094937 A1, 28 April 2011.
126. Mudraboyina, B.P.; Fu, D.; Jessop, P.G. Supercritical fluid rectification of lignin microwave-pyrolysis oil. *Green Chem.* **2015**, *17*, 169–172. [CrossRef]
127. Montesantos, N.; Pedersen, T.H.; Nielsen, R.P.; Rosendahl, L.A.; Maschietti, M. High-temperature extraction of lignocellulosic bio-crude by supercritical carbon dioxide. *Chem. Eng. Trans.* **2019**, *74*, 799–804.
128. Chan, Y.H.; Yusup, S.; Quitain, A.T.; Chai, Y.H.; Uemura, Y.; Loh, S.K. Extraction of palm kernel shell derived pyrolysis oil by supercritical carbon dioxide: Evaluation and modeling of phenol solubility. *Biomass Bioenergy* **2018**, *116*, 106–112. [CrossRef]
129. Chan, Y.H.; Yusup, S.; Quitain, A.T.; Uemura, Y.; Loh, S.K. Fractionation of pyrolysis oil via supercritical carbon dioxide extraction: Optimization study using response surface methodology (RSM). *Biomass Bioenergy* **2017**, *107*, 155–163. [CrossRef]
130. Feng, Y.; Meier, D. Comparison of supercritical CO_2, liquid CO_2, and solvent extraction of chemicals from a commercial slow pyrolysis liquid of beech wood. *Biomass Bioenergy* **2016**, *85*, 346–354. [CrossRef]
131. Cheng, T.; Han, Y.; Zhang, Y.; Xu, C. Molecular composition of oxygenated compounds in fast pyrolysis bio-oil and its supercritical fluid extracts. *Fuel* **2016**, *172*, 49–57. [CrossRef]
132. Patel, R.N.; Bandyopadhyay, S.; Ganesh, A. Extraction of cardanol and phenol from bio-oils obtained through vacuum pyrolysis of biomass using supercritical fluid extraction. *Energy* **2011**, *36*, 1535–1542. [CrossRef]
133. Rout, P.K.; Naik, M.K.; Naik, S.N.; Goud, V.V.; Das, L.M.; Dalai, A.K. Supercritical CO_2 fractionation of bio-oil produced from mixed biomass of wheat and wood sawdust. *Energy Fuels* **2009**, *23*, 6181–6188. [CrossRef]
134. Wang, J.; Cui, H.; Wei, S.; Zhuo, S.; Wang, L.; Li, Z.; Yi, W. Separation of biomass pyrolysis oil by supercritical CO_2 extraction. *Smart Grid Renew. Energy* **2010**, *01*, 98–107. [CrossRef]
135. ISO 8217:2017. *Petroleum Products-Fuels (class F)-Specifications of Marine Fuels*; ISO: Geneva, Switzerland, 2017.
136. Pedersen, K.S.; Christensen, P.L.; Shaikh, J.A. *Phase Behavior of Petroleum Reservoir Fluids*, 2nd ed.; CRC Press: Boca Raton, FL, USA, 2014; ISBN 9780429110306.
137. Chrastil, J. Solubility of solids and liquids in supercritical gases. *J. Phys. Chem.* **1982**, *86*, 3016–3021. [CrossRef]
138. Dohrn, R.; Brunner, G. High-pressure fluid-phase equilibria: Experimental methods and systems investigated (1988–1993). *Fluid Phase Equilib.* **1995**, *106*, 213–282. [CrossRef]

139. Laugier, S.; Richon, D. High-pressure vapor-liquid equilibria of two binary systems: Carbon dioxide + cyclohexanol and carbon dioxide + cyclohexanone. *J. Chem. Eng. Data* **1997**, *42*, 155–159. [CrossRef]
140. Jing, Y.; Hou, Y.; Wu, W.; Liu, W.; Zhang, B. Solubility of 5-Hydroxymethylfurfural in supercritical carbon dioxide with and without ethanol as cosolvent at (314.1 to 343.2) K. *J. Chem. Eng. Data* **2011**, *56*, 298–302. [CrossRef]
141. Sparks, D.L.; Hernandez, R.; Estévez, L.A.; Holmes, W.E.; French, W.T. Solubility of small-chain fatty acids in supercritical carbon dioxide. *AIChE Annu. Meet. Conf. Proc.* **2008**, *55*, 4922–4927.
142. Bamberger, T.; Erickson, J.C.; Cooney, C.L.; Kumar, S.K. Measurement and model prediction of solubilities of pure fatty acids, pure triglycerides, and mixtures of triglycerides in supercritical carbon dioxide. *J. Chem. Eng. Data* **1988**, *33*, 327–333. [CrossRef]
143. Maheshwari, P.; Nikolov, Z.L.; White, T.M.; Hartel, R. Solubility of fatty acids in supercritical carbon dioxide. *J. Am. Oil Chem. Soc.* **1992**, *69*, 1069–1076. [CrossRef]
144. Iwai, Y.; Fukuda, T.; Koga, Y.; Arai, Y. Solubilities of myristic acid, palmitic acid, and cetyl alcohol in supercritical carbon dioxide at 35 °C. *J. Chem. Eng. Data* **1991**, *36*, 430–432. [CrossRef]
145. Garlapati, C.; Madras, G. Solubilities of palmitic and stearic fatty acids in supercritical carbon dioxide. *J. Chem. Thermodyn.* **2010**, *42*, 193–197. [CrossRef]
146. Brandt, L.; Elizalde-Solis, O.; Galicia-Luna, L.A.; Gmehling, J. Solubility and density measurements of palmitic acid in supercritical carbon dioxide + alcohol mixtures. *Fluid Phase Equilib.* **2010**, *289*, 72–79. [CrossRef]
147. Schwarz, C.E.; Knoetze, J.H. Phase equilibrium measurements of long chain acids in supercritical carbon dioxide. *J. Supercrit. Fluids* **2012**, *66*, 36–48. [CrossRef]
148. Bamberger, A.; Schmelzer, J.; Walther, D.; Maurer, G. High-pressure vapour-liquid equilibria in binary mixtures of carbon dioxide and benzene compounds: experimental data for mixtures with ethylbenzene, isopropylbenzene, 1,2,4-trimethylbenzene, 1,3,5-trimethylbenzene, ethenylbenzene and isopropenylbenzene, and their correlation with the generalized Bender and Skjold-Jorgensen's group contribution equation of state. *Fluid Phase Equilib.* **1994**, *97*, 167–189. [CrossRef]
149. Jennings, D.W.; Schucker, R.C. Comparison of high-pressure vapor-liquid equilibria of mixtures of CO_2 or propane with nonane and C9 alkylbenzenes. *J. Chem. Eng. Data* **1996**, *41*, 831–838. [CrossRef]
150. Phiong, H.S.; Lucien, F.P. Volumetric expansion and vapour-liquid equilibria of α-methylstyrene and cumene with carbon dioxide at elevated pressure. *J. Supercrit. Fluids* **2003**, *25*, 99–107. [CrossRef]
151. McHugh, M.; Paulaitls, M.E. Solid solubilities of naphthalene and biphenyl in supercritical carbon dioxide. *J. Chem. Eng. Data* **1980**, *25*, 326–329. [CrossRef]
152. Yanagiuchi, M.; Ueda, T.; Matsubara, K.; Inomata, H.; Arai, K.; Saito, S. Fundamental investigation on supercritical extraction of coal-derived aromatic compounds. *J. Supercrit. Fluids* **1991**, *4*, 145–151. [CrossRef]
153. Wells, P.A.; Chaplin, R.P.; Foster, N.R. Solubility of phenylacetic acid and vanillan in supercritical carbon dioxide. *J. Supercrit. Fluids* **1990**, *3*, 8–14. [CrossRef]
154. Schmitt, W.J.; Reid, R.C. Solubility of monofunctional organic solids in chemically diverse supercritical fluids. *J. Chem. Eng. Data* **1986**, *31*, 204–212. [CrossRef]
155. Kurnlk, R.T.; Holla, S.J.; Reid, R.C. Solubility of solids in supercritical carbon dioxide and ethylene. *J. Chem. Eng. Data* **1981**, *26*, 47–51. [CrossRef]
156. García-González, J.; Molina, M.J.; Rodríguez, F.; Mirada, F. Solubilities of phenol and pyrocatechol in supercritical carbon dioxide. *J. Chem. Eng. Data* **2001**, *46*, 918–921. [CrossRef]
157. Van Leer, R.A.; Paulaitis, M.E. Solubilities of phenol and chlorinated phenols in supercritical carbon dioxide. *J. Chem. Eng. Data* **1980**, *25*, 257–259. [CrossRef]
158. Pfohl, O.; Brunner, G. Two-and three-phase equilibria in systems containing benzene derivatives, carbon dioxide, and water at 373.15 K and 10–30 MPa. *Fluid Phase Equilibria* **1997**, *141*, 179–206. [CrossRef]
159. Yamini, Y.; Fat'Hi, M.R.; Alizadeh, N.; Shamsipur, M. Solubility of dihydroxybenzene isomers in supercritical carbon dioxide. *Fluid Phase Equilibria* **1998**, *152*, 299–305. [CrossRef]
160. Škerget, M.; Čretnik, L.; Knez, Ž.; Škrinjar, M. Influence of the aromatic ring substituents on phase equilibria of vanillins in binary systems with CO_2. *Fluid Phase Equilibria* **2005**, *231*, 11–19. [CrossRef]
161. Liu, J.; Kim, Y.; McHugh, M.A. Phase behavior of the vanillin-CO_2 system at high pressures. *J. Supercrit. Fluids* **2006**, *39*, 201–205. [CrossRef]

162. Lee, M.J.; Kou, C.F.; Cheng, J.W.; Lin, H.M. Vapor-liquid equilibria for binary mixtures of carbon dioxide with 1,2-dimethoxybenzene, 2-methoxyphenol, or p-cresol at elevated pressures. *Fluid Phase Equilibria* **1999**, *162*, 211–224. [CrossRef]
163. Pfohl, O.; Pagel, A.; Brunner, G. Phase equilibria in systems containing o-cresol, p-cresol, carbon dioxide, and ethanol at 323.15-473.15 K and 10-35 MPa. *Fluid Phase Equilibria* **1999**, *157*, 53–79. [CrossRef]
164. Lee, R.J.; Chao, K.C. Extraction of 1-methylnaphthalene and m-cresol with supercritical carbon dioxide and ethane. *Fluid Phase Equilibria* **1988**, *43*, 329–340. [CrossRef]
165. Kambo, H.S.; Dutta, A. A comparative review of biochar and hydrochar in terms of production, physico-chemical properties and applications. *Renew. Sustain. Energy Rev.* **2015**, *45*, 359–378. [CrossRef]

© 2020 by the authors. Licensee MDPI, Basel, Switzerland. This article is an open access article distributed under the terms and conditions of the Creative Commons Attribution (CC BY) license (http://creativecommons.org/licenses/by/4.0/).

Article

Decision-Making Process in the Circular Economy: A Case Study on University Food Waste-to-Energy Actions in Latin America

Laura Brenes-Peralta [1], María F. Jiménez-Morales [2], Rooel Campos-Rodríguez [2], Fabio De Menna [1,*] and Matteo Vittuari [1]

1. Department of Agricultural and Food Sciences, University of Bologna, 40127 Bologna, Italy; laura.brenesperalta2@unibo.it (L.B.-P.); matteo.vittuari@unibo.it (M.V.)
2. Agribusiness School, Tecnológico de Costa Rica, Cartago 30101, Costa Rica; maria.jimenez@tec.ac.cr (M.F.J.-M.); rocampos@tec.ac.cr (R.C.-R.)
* Correspondence: fabio.demenna2@unibo.it

Received: 2 April 2020; Accepted: 25 April 2020; Published: 6 May 2020

Abstract: Economies have begun to shift from linear to circular, adopting, among others, waste-to-energy approaches. Waste management is known to be a paramount challenge, and food waste (FW) in particular, has gained the interest of several actors due to its potential impacts and energy recovery opportunities. However, the selection of alternative valorization scenarios can pose several queries in certain contexts. This paper evaluates four FW valorization scenarios based on anaerobic digestion and composting, in comparison to landfilling, by applying a consistent decision-making framework through a combination of linear programming, Life Cycle Thinking (LCT), and Analytic Hierarchy Process (AHP). The evaluation was built upon a case study of five universities in Costa Rica and portrayed the trade-offs between environmental impacts and cost categories from the scenarios and their side flows. Results indicate that the landfill scenario entails higher Global Warming Potential and Fresh Water Eutrophication impacts than the valorization scenarios; however, other impact categories and costs are affected. Centralized recovery facilities can increase the Global Warming Potential and the Land Use compared to semi-centralized ones. Experts provided insights, regarding the ease of adoption of composting, in contrast to the potential of energy sources substitution and economic savings from anaerobic digestion.

Keywords: centralized waste valorization; lifecycle thinking; AHP; side flow; anaerobic digestion; composting

---

1. Introduction

The circular economy is regarded as a sustainable economic system that reduces raw material extraction and recirculates resources, while creating benefits to society, industries, and the environment [1]. Strategies for its achievement include principles from various schools of thought, where designing out what is commonly observed as waste is fundamental [2] because of the value remaining in these materials. Waste, together with productive activities and transport, are relevant sources of environmental degradation and impacts, such as global warming, which involves a significant risk for humanity [3]. Ordinary waste entails almost 50% of organic sources approximately, and food waste (FW) is the highest contributor of that organic fraction [4]. Globally, it is accountable for 4.4 Gt CO_2 eq per year [5]. Consequently, the disregard of FW causes economic, social, and environmental constraints [6,7].

Target 12.3 of the Sustainable Development Goals (SDG) aims to halve FW by 2030 [8], and in pursuit of that reduction, the "food use-not-waste" hierarchy embraces alternatives from most to less desirable, similar to the waste management hierarchy from the EU Waste Framework Directive 2008/98/EC [9,10]. FW management begins with the prevention and optimization of food use and supply. Actions include the avoidance of FW throughout the supply chain, as well as the donation and redistribution of surplus for human consumption when possible, and then the allocation to animal feed or non-food product transformation. Once FW occurs, it shall be valorized and treated through recycling and energy recovery, before landfilling [9–11]. That final disposal is commonly perceived as the least preferable option, due to its high environmental implications, such as emissions to soil, water, and air, occurring during biowaste degradation [10].

The interest in the circular economy and FW reduction in food systems rests, among other reasons, on the fact that this sector is high in energy demand, and suffers from intake-output energy imbalances [12]. Therefore, the recovery of currently wasted energy embodied in FW can represent an opportunity to recirculate energy into human activities again, thus aiding into more sustainable systems.

Depending on FW composition, anaerobic digestion and composting, are regarded as suitable options for energy recovery [12]. Anaerobic Digestion (AD) consists of the anaerobic degradation of the residues while generating biogas and digestate, which can be used as fertilizer with lower environmental impacts [13–16]. Even when suggesting the fittingness of the obtained by-products from this alternative, various authors recommend a close observation in regards to the source and composition of the FW and co-digesting materials, and the technical and economic potential challenges [17–20]. Composting (CP) is defined as the controlled organic waste degradation through biological agents, suitable to treat the biological fraction of ordinary waste [10], resulting in a rich soil substrate. Experiences using the Takakura composting method has proven it to be an efficient option for food and garden biowaste treatment [21], while remaining a relatively easy-to-adopt practice at domestic or larger scales [22,23].

Even when preferred over landfilling, FW valorization will also have recognized risks and embedded effects, due to emissions, transport, degradation, and labor [22,24]. Therefore, decision-making processes to support the selection of one option or another are not simple. Life Cycle Thinking (LCT) is considered to be an apt approach to evaluate food waste valorization alternatives through methods like Life Cycle Assessment (LCA) and Environmental Life Cycle Costing (E-LCC) [7,25–27]. In addition, multicriteria decision methods can aid managers and policy-makers from different levels to analyze the trade-offs offered by science-based evidence from the evaluation of different alternatives [28].

Several studies in certain regions such as Latin America and the Caribbean, focus on waste generation and composition analysis [4,29–32] and few of them directly involve LCT [33] or decision-making approaches for waste management, neither a combination of those. Therefore, gaps in literature availability and decisional frameworks, suggest integrated approaches are required [34]. Enormous amounts of biomass are possibly available for circular strategies in this Region, since 54% of the 160 million tons of its yearly waste belonging to biowaste, is generally disposed in landfills [31]. Even when a regional agreement or framework for bioeconomy is lacking, Latin American countries have recently begun to undertake specific policies towards a circular economy and food waste valorization [35]. Costa Rica, in particular, launched several initiatives on this matter, such as the inter-sectorial actions led by the Costa Rican Food Loss and Waste Network [36], the National Policy on Sustainable Production and Consumption [37], the National Decarbonisation Strategy [38], and the Integrated Waste Management Law no.8839 [39]. On one hand, these policies motivate different stakeholders to pursue FW valorization actions; on the other, it enables actions that would directly support further steps into the achievement of the SDGs, such as less FW generation and waste management alternatives with lower emissions.

This paper evaluates FW valorization alternatives and compares them to the business-as-usual FW landfilling, through a combination of methods that includes linear programming to determine an optimal collection route for the waste, environmental and economic potential impacts analysis through a system-expanded LCA and E-LCC, and the prioritization of alternatives through an Analytic Hierarchy Process (AHP). The evaluation was built upon a case study of five universities, and it is one of the first assessments of this kind in Costa Rica and the Latin American region. The final aim is to contribute to decision-making processes to move into more circular approaches at the university consortium level, but also at the local level by offering a consistent framework to support actions to valorize FW. Potentially, the study can help other similar institutions, small communities or even small municipalities to plan for their biodegradable waste management in small centralized or semi-centralized units, and prioritize sustainable approaches to address food waste.

2. Materials and Methods

2.1. Methodological Framework and Case Description

This case study proposed a decision-making process for food waste-to-energy scenarios, through their evaluation and comparison to a business-as-usual scenario, landfill (LF).

It aggregated the FW from a consortium of five universities located in and nearby the Central Valley of Costa Rica, belonging to a national network of sustainable education institutions called REDIES. Rojas-Vargas et al. [30] determined the amount of FW generated in these university canteens using the standardized guidelines to measure FW in restaurants provided by the Costa Rican Food Loss and Waste Network [40]. That first and only available formal study on FW quantification for a group of universities in the country amounted 2.607 tonnes of FW per week, with an operative service of 45 weeks, given their academic calendar. There are different food waste definitions; and this paper adopts the FW conceptualization reported by the FUSIONS definitional framework that describes it as "any food, and inedible parts of food, removed from the food supply chain to be recovered or disposed" [41].

Being the landfilling disposal a common practice in Costa Rica, and generally in Latin America, this study proposed to follow the "food use-not-waste" hierarchy [9], which can be easily aligned to the Costa Rican Waste Management Law [39] and the REFRESH Generic strategy for LCA and LCC [42]. This study proposed to move from a situation where FW was disposed at the landfill, to a situation of valorization, or food waste-to-energy alternatives. This perspective was similar to the one described by the REFRESH strategy as REFRESH Situation (RS) RS 4 to RS 3, since the university consortium would agree to hand over the FW for valorization as part of their waste management (RS 3) instead of sending it to an end-of-life treatment or landfill (RS 4) as presented in Figure 1.

Anaerobic Digestion (AD) and Composting (CP) are among the alternatives to be considered, generating a side flow that has some value with the potential to replace a product on the market.

The overall methodological framework, accompanied by an iterative literature review, combined three methods (Figure 2) in a step-wise sequence.

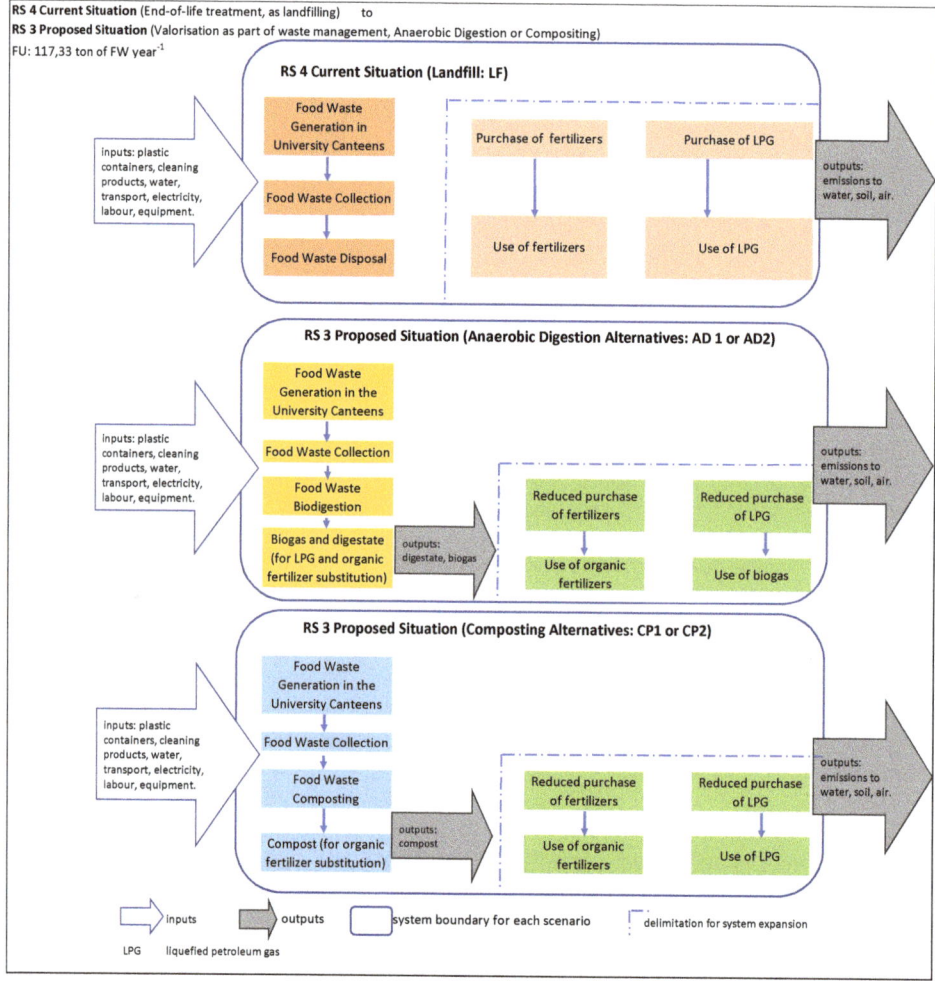

Figure 1. Flow diagram of proposed situations in the University Consortium, for valorization through FW-to-energy alternatives. Source: [42] adapted by the authors.

The first method aimed to define the FW valorization sites and collection routes, through Linear programming. This led to two possible route designs that were used to model four FW valorization scenarios. Once the scenarios were defined and supported by literature reviews and by experts, a pre-selection of evaluation criteria was considered to later conduct a system-expanded LCA and E-LCC, which considered the impacts caused by the valorization scenarios, as well as the avoided impacts since FW would be diverted from LF and side flows would be utilized. LCA and E-LCC allowed to observe the performance of each scenario in terms of the environmental impacts and costs categories, offering relevant data for an experts' assessment developed through the third applied method of this framework: the AHP. This latter allowed to prioritize the scenarios within the local context and following a science-to-expert approach.

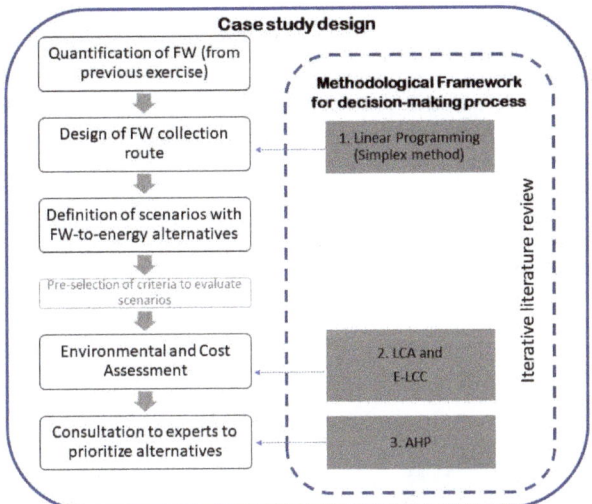

Figure 2. Methodological framework for a decision-making process of food waste (FW) valorization alternatives.

2.2. Route Optimization

Since the five universities have accredited environmental managers and operate under the university autonomy principle, it was assumed that they could potentially treat the FW and would agree on diverting the FW from a business-as-usual to a valorization scenario. Therefore, a simulation of an FW collection route was performed for this consortium. First, the FW valorization plant location was evaluated, through a decision matrix defined by the researchers, with four criteria and the following qualification scale and weight, based on similar techniques presented by various authors [43,44]:

1. Space availability to install the waste valorization plant: this criterium would receive a binomial response, where a value of 1 will be assigned if the campus had an available area of at least 250 m^2 where a waste valorization plant can be established without negatively affecting the university activities, 0 would be given if that space was not available. This area (m^2) was selected base on the experience and observation of biowaste treatment facilities in municipalities and institutions;
2. Technical capacity to operate an FW treatment process: with a binomial response also, the site would receive a score of 1 if there was a minimum of one professional at campus capable and knowledgeable in waste management at least at the pilot scale, 0 if that kind of professional was unavailable;
3. Available infrastructure: the binomial qualification for this criterion would consist of assigning a value of 1 if the campus had at least one operative anaerobic digestor or composter to process the FW, 0 if they did not have any infrastructure for FW treatment;
4. FW quantity: a 5-value scale was assigned in this criterion, where 5 corresponds to the highest FW quantity generated within the group of campuses, and 1 for the smallest amount of FW.

Each criterion was weighted [43–46] from a full score of 100%, as follows: a weight of 15% was assigned to technical capacity, supported by a law requirement in the country, a 25% weight was assigned to space availability and FW quantity, since they would have a higher impact than the previous criterion but in equal conditions among themselves. Finally, a 35% weight was assigned to available infrastructure, since it would have higher importance than the previous in terms of the possibility of short-term establishment of the valorization alternatives and budget implications. The location(s) with the highest score were to be selected to install the FW valorization facility since it would have available space, technical capacities, available infrastructure and higher amounts of FW to process.

Afterwards, the researchers calculated the average distances between each FW generation site (institutions) using Google maps. This allowed us to obtain the distance in kilometers between each two points, later used in the route design, with the assumption that budget constraints in the universities would only allow one truck for a weekly FW transportation. The five institutions were codified as A, B, C, D and E, corresponding to the five campuses of this study (Figure 3). A value matrix was set up with the average distances between each two points (Table 1) and modeled by linear programming, using the Simplex LP Method [47] and the Solver Tool from Microsoft ® Excel ® (2019 MSO Version, Microsoft Corporation ©, Redmond, WA, USA), to obtain the optimal route by minimizing the total distance.

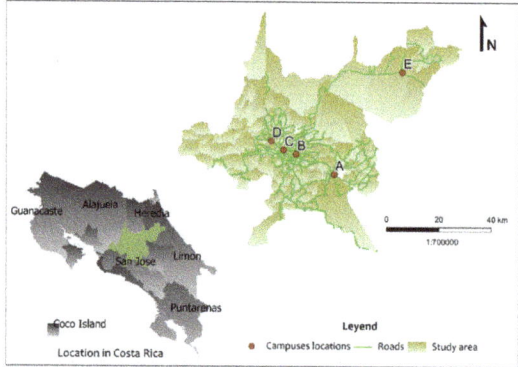

Figure 3. Location of the five campuses of the universities from the consortium (image developed by Mariajosé Esquivel, using Costa Rican Digital Atlas, 2014. Projection: CRTM05. Q GIS Software, Version V3.4, The Open Source Geospatial Foundation OSGeo, Chicago, IL, USA).

Table 1. Distance matrix among the five university campuses (sites) in kilometers (km).

SITES	A	B	C	D	E
A	0	22	40	34	104
B	22	0	16	11	86
C	40	16	0	11	82
D	34	11	11	0	95
E	104	86	82	95	0

Due to the farther distance from location E to the rest of the group, a second linear programming model was calculated (Table 2).

Table 2. Distance matrix among four of the university campuses (sites) in kilometers (km).

SITES	A	B	C	D
A	0	22	40	34
B	22	0	16	11
C	40	16	0	11
D	34	11	11	0

Two routes were calculated, one that would accept all the FW from the first four generation sites and deliver it at the fifth campus for centralized valorization (coded as 1), and a second route that would consist of a semi-centralized valorization where more than one campus would be in charge of processing the FW (coded as 2). The obtained data was later used in the LCA and E-LCC to compare the effect of the two routes on each FW valorization alternatives.

2.3. LCA and LCC

2.3.1. Goal and Scope

Following the ISO14040 Standard [48] and Hunkeler D., Lichtenvort K., and Rebitzer, G. [49] respectively, LCA and E-LCC were used to understand the environmental and economic effects the consortium of universities would have as a result of moving from a business-as-usual to an FW valorization scenario. This consortium with already well-established FW measurement and environmental management units, defined the system boundaries from gate to gate: from the FW generation point to the campus where the valorization facilities would be established and side flows would be obtained [42,50]. These side flows, have an already existing market value in Costa Rica [51] and could be used by the same university or a third party, who would collect them at the campus gate.

2.3.2. Reference Flows and Functional Unit

The study uses a reference flow that consists of a mass-based unit for the LCA and monetary-based units for the E-LCC, considering as functional unit (FU) the amount of treated FW per year: 117.3 t of FW per year.

2.3.3. Environmental Impact Categories, Cost Elements and Assessment Methods

Literature reviews, the criteria of the researchers and a set of three advisors with international, regional and national experience in FW and waste management suggested the main indicators to evaluate the alternatives. The two main environmental impacts were Gobal Warming Potential and Land-Use, both consistent with the recommended categories in FW LCA analysis, as well as with Costa Rican aim on decarbonization. Midpoint indicators were preferred by this study in order to observe particular impact categories for this type of FW valorization processes. Therefore, the ReCiPe 2016 midpoint method, Hierarchic version (developed by RIVM, Radboud University Nijmegen, Leiden University and PRé Sustainability) was applied using SimaPro (Version 9.0.0.49, PRè Consultants ©, Amersfoort, The Netherlands). In general, the study calculated these potential environmental impacts, focusing mostly on the first two:

Global Warming Potential (GWP), expressed in kg CO_2 eq;
Land-Use (LU), expressed in m^2a crop;
Terrestrial Acidification (TA), expressed in kg SO_2 eq;
Freshwater Eutrophication (FE), expressed in kg P eq;
Mineral Resource Scarcity (M-RS), expressed in kg Cu eq;
Fossil Resource Scarcity (F-RS), expressed in kg oil eq;
Water Consumption (WC), expressed in m^3.

The E-LCC included the following cost categories: inputs, labor, transport, public services and depreciation from equipment investments. They were categorized in four main groups: a-inputs and labor at the generation point, b-transport to the disposal or valorization site, c-valorization system (this one includes all operation elements such as inputs, labor, energy, water), and d-depreciation due to the use of the equipment in which the consortium shall invest. The depreciation cost related to the required investment and the net economic effect as a result of the overall operative costs and savings during the valorisation of the FW, were selected as indicators in the economic dimension, expressed in American dollars (USD).

A category of social-oriented indicators, such as job generation and ease-of-implementation were considered as well. Job generation was calculated after the FW valorization labour requirement was inventoried for the E-LCC and then translated into the amount of new required full-time collaborators for each scenario. The ease-of-implementation was defined as the attribute that expresses how practical or less complex an alternative was in terms of technique, equipment and operation. Both indicators were assessed by the experts during the AHP implementation. Therefore, these indicators not only guided the proper data collection in the inventory phase of the LCA and E-LCC, but were the ones to be considered as criteria during the science-to-expert approach.

2.3.4. Scenarios and Inventory

The alternatives consisted of Anaerobic Digestion (AD) and Composting (CP) as seen in Figure 1 (more detailed information in the Supplementary Materials). The assessment comprised four FW valorization scenarios:

AD1: Anaerobic Digestion in a centralized plant and FW collection route design 1. It proposed the operation of a continuous-load digester, and considered there was an already existing and operational digester on the selected site that would have the capacity to process the annual amount of FW.

AD2: Anaerobic Digestion in a semi-centralized alternative of three valorization plants with route design 2. This scenario required three continuous-load digesters, one was already operating in one site, and new digesters would have to be set in two more valorization sites. It is assumed that the digestate from the already operative digestor would help to establish the microbiota in the other two locations.

CP1: Composting in a centralized plant and FW collection route design 1. This scenario anticipated a modified and scaled Takakura composting method, operated through a set of seven automatic composters (the Model JK5100 ®, Joraform, Laholm, Sweden) available in the market to manage 0.08 ton of FW per day each.

CP2: Composting in a semi-centralized alternative of three valorization plants with an FW collection route design 2. This scenario would also use a modified and scaled Takakura composting method, operated through a set of six new JK5100 ® automatic composters capable to process 0.08 tonnes of FW per day each and one already similar composter in one of the sites.

In the business-as-usual scenario, the FW was collected and disposed of in a landfill (LF) by an authorized third party. It was modeled upon national data regarding FW collection and disposal costs [52,53], and calculated distances in Google Maps, from the campus to the closest landfill where ordinary wastes would be usually directed to, according to the Environmental managers of these institutions.

An inventory of the inputs and outputs of each scenario was performed, with data gathered from previous experiments [22], literature and a questionnaire filled by the restaurant manager and operators. When necessary, the allocation of certain inputs based on the FW generation proportion of each campus was applied, due to a lack of primary data in some of the institutions. Inputs consisted of plastic containers to collect FW, products and water to clean (both the generation and valorization sites), transport of those inputs, and electricity to pre-condition FW, as well as the required labor to operate each stage. The FW transport was calculated regarding the FW mobilized mass and the FW transportation route; this meant that it considered the kilometers in the business-as-usual route for LF, as well as the kilometers for route 1 or route 2 obtained in the optimized route design. Outputs included the compost or digestate, biogas depending on the valorization alternative, the wastewater, as well as the correspondent emissions for the valorization process and expanded system. Packaging waste from inputs were not included since they would be considered to be outside the system boundaries of the present study. Finally, processes for the correspondent FW treatment or disposition were selected from the Ecoinvent database, whether it was a landfill for municipal solid waste, biowaste anaerobic digestion, or industrial composting on each scenario.

2.3.5. Assumptions and Data Sources

Several literatures present AD and CP as alternatives for FW valorization, and prior studies in one of the universities concluded that those were technically fit within the local conditions, which motivated further analysis and the assumption that these would be the valorization alternatives to be assessed in this study. Most inputs were considered as yearly consumables, except for the plastic containers for the FW, which were estimated to have a life of five years; therefore, the cost and mass for the yearly FW treatment was estimated. All alternative scenarios suppose that 50% of water consumption in cleaning operations would come from rainwater collection, a practice that is becoming more usual in the country. The compost yield was estimated to be 18.75% from the mass of the FW [22]. The biogas yield was obtained from literature reviews regarding biogas production, digestate production, technical characteristics, and calorific potentials, to assume a methane production of 53% of the produced biogas [13,54–57]. Distances from input suppliers as well as from the FW generation to valorization sites were calculated with Google Maps, and databases like Ecoinvent 3.4 were used for the inventoried processes on each scenario. The exchange rate to convert Costa Rican market prices (CRC) into American dollars (USD) was retrieved from the Costa Rican Central Bank at the moment of the study, at a rate of 596.18 CRC: 1 USD. Other information sources included scientific literature, environmental declarations, Costa Rican public services databases and market prices.

2.3.6. Interpretation

Critical stages or hotspots regarding environmental and economic data were identified in the business-as-usual and alternative scenarios, and an evaluation regarding the avoidance of certain impacts through a system expansion was used for a comparison among the four alternatives. A sensitivity analysis was conducted to observe the result of potential input changes suggested by experts during the exercise as well as contextual conditions. The summary of the LCA and E-LCC results was presented to a group of experts, to prioritize the option with more potential to be adopted by the consortium.

2.4. Multicriteria Decision Method: AHP

Saaty (2008) established the basis of a multicriteria decision-making approach named Analytic Hierarchy Process (AHP) [28], in which factors are arranged in a hierarchic structure. It follows a systematic set of steps, beginning with the definition of the problem. The second step sets a decision hierarchy structure where the goal of the process is placed at the first level of the structure, criteria on which subsequent elements depend are placed at the intermediate level and the alternatives to be considered in the decision process rest at the lowest level. A third step consists of the construction of pairwise comparison matrices, which are later normalized and an eigenvector is determined to later, in the fourth step, use the obtained priority vectors in the pairwise comparison to weigh the alternatives in the subsequent level. In this study, the goal was to select an FW valorization alternative for this University Consortium, based on pairwise comparisons of environmental, economic and social criteria to later prioritize the FW valorization alternatives.

This study considered six criteria from the environmental, economic and social dimensions (Figure 4), regarded as Global Warming Potential, Land-Use, depreciation cost (linked to the required investment), net economic effect, ease-of-implementation and job generation.

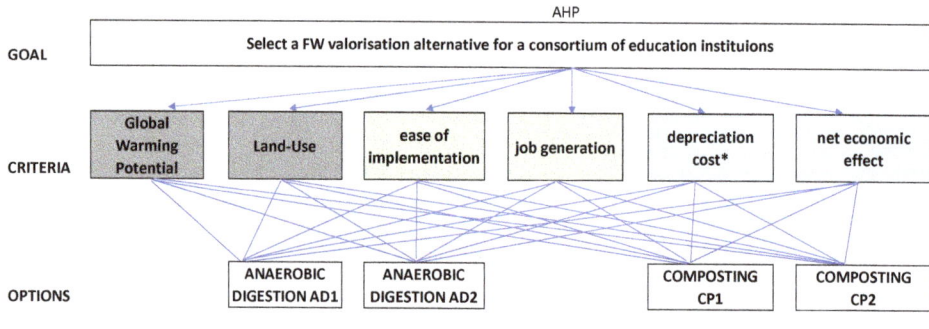

Figure 4. Analytic Hierarchy Process (AHP) structure to evaluate FW valorization alternatives for the university consortium.

A group of 10 experts was reached to provide a science-to-expert approach. The technical information from the description of the scenarios and the results of the LCA and E-LCC was offered to these professionals. The group of experts was gender-balanced and formed by professionals in environmental sciences, engineering and economics, most of them with postgraduate education and currently holding positions as environmental managers from education institutions or local governments, policy makers, academics, specialists in international/non-governmental organizations, or waste valorization entrepreneurs. The responses of the experts where registered in a data collection tool created for this purpose (validated and tested, Cronbach's α = 0.93).

They were asked first to prioritize each criterion versus the other and then to judge each FW-to-energy alternative versus the other regarding the mentioned criteria, using the fundamental scale for absolute numbers. This nine-point scale qualifies the alternatives in regards of the intensity of importance of one option over the other, assigning a value of 1 when there is equal importance, and up to a value of 9 when there is extreme importance [28]. The results were aggregated by a geometric mean, computed into a matrix, and later normalized using Microsoft ® Excel ®. The Eigenvector was calculated and used to weight the results for the four valorization alternatives. Then, the process allowed to prioritize the FW-to-energy or valorization alternatives for the consortium. A consistency check was performed for each matrix through the calculation of the Consistency Ratio (CR ≤ 0.10). Finally, experts observed the overall results and offered feedback on the methodological framework in a single open-question section of the data collection tool.

3. Results

3.1. Selection of FW Valorization Facility and Route Optimization

The five campuses were evaluated to assess the possibility of establishing an FW valorization plant (Table 3).

Table 3. Evaluation matrix for the selection of the FW treatment site.

Site	Available Space	Technical Capacity	Available Infrastructure	FW Quantity	Score
A	25	15	0	25	65
B	0	15	0	5	20
C	25	15	35	15	90
D	0	15	0	10	25
E	25	15	35	20	95

As part of the design of the scenarios, this study qualified the capacity of the five campuses to implement FW valorization alternatives. Sites A, C and E had available space, and C and E would have already existing infrastructure and equipment for at least one of the valorization alternatives. Site A obtained the highest score due to the generation of 41.72 ton of FW per year, followed by E (26.51 ton FW year^{-1}), C (20.33 ton FW year^{-1}), B (20.32 ton FW year^{-1}) and D (8.45 ton FW year^{-1}). Consequently, after assigning the values and weight for each criterion, site E was defined as the site of preference to establish an FW valorization plant. Sites C and A would follow, one because of the existence of space and infrastructure, and the other because of the available space and amount of FW which could remain in place in order to avoid its transportation around the consortium. Sites B and D had limited capacities for the establishment of FW-to-energy alternatives.

Considering those results, there was a fist calculation of the best possible route design, named route 1. It consisted of 126 km and FW would be collected first at site A, and continue to point B, to D, to C and finalize in E where the valorization would take place.

The second route calculation proposed a collection route with four sites. In that case, the FW collection route 2 would consist of 33 km of FW transportation. It entailed a semi-centralized valorisation system, where the sites that obtained the second and third highest scores in the site evaluation matrix for the plant selection (Table 3) would become FW treatment facilities as well. Therefore, FW from site B would be transported to site A, where FW valorization of both sites would take place; in parallel, site C would valorize its own FW and the one carried from site D; and site E would process its own waste. It would be still done with the restriction of one single truck for FW collection.

3.2. LCA and E-LCC

The second method of this framework, based on LCT allowed us to observe the different environmental and cost impacts of the evaluated scenarios.

Table 4 presents the LCA results, regarding the impact categories selected for this study, and from which the GWP and LU were considered for the further science-to-expert approach.

Table 4. Life Cycle Assessment (LCA) for the business as usual and alternative FW disposal or treatment scenarios.

Impact Category	Unit	LF	AD1	AD2	CP1	CP2
Global Warming	kg CO_2 eq	90,050.00	16,113.16	11,906.42	13,973.26	9376.31
Terrestrial Acidification	kg SO_2 eq	17.01	40.11	25.45	193.92	177.66
Freshwater Eutrophication	kg P eq	2.23	2.20	1.92	1.80	1.34
Land Use	m^2a crop eq	388.48	230.30	132.64	542.52	408.88
Mineral Resource Scarcity	kg Cu eq	6.84	16.80	11.68	29.92	16.41
Fossil Resource Scarcit	kg oil eq	1140.30	3034.22	1553.35	2992.81	1417.40
Water Consumption	m^3	58.22	59.74	88.57	55.53	80.32

The disposal of the FW in a business-as-usual scenario such as LF, presents higher Global Warming Potential and Freshwater Eutrophication impacts than the four valorization scenarios. However, CP1 and CP2 present a higher Land-Use than the rest, while AD1 and AD2, has the lowest Land-Use impact of the scenarios. LF has lower Acidification Potential than scenarios AD1, CP1 and CP2; and the four alternative scenarios would have higher Mineral Resources and Fossil Resources depletion than LF. Water Consumption is also increased in all valorization alternatives, except CP1.

Figures 5–12 summarise the environmental impacts, detailed in three phases or stages for each scenario: inputs, FW transport and FW disposal (or treatment).

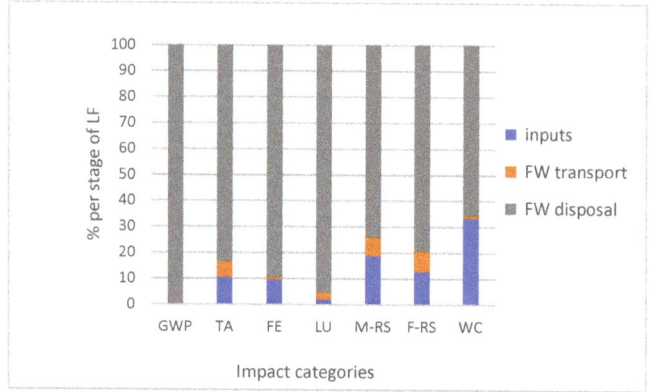

Figure 5. Impacts from FW landfilling disposal (LF).

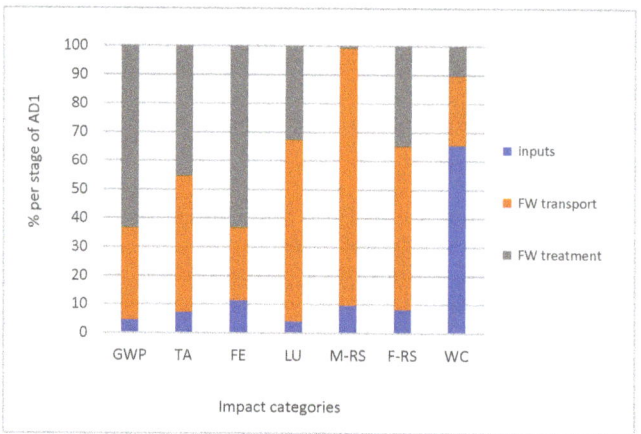

Figure 6. Impacts from FW Anaerobic Digestion in a centralized scenario (AD1).

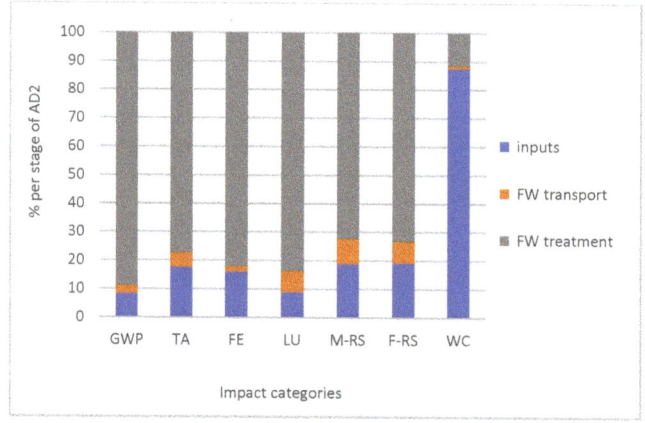

Figure 7. Impacts from FW Anaerobic Digestion in a semi-centralized scenario (AD2).

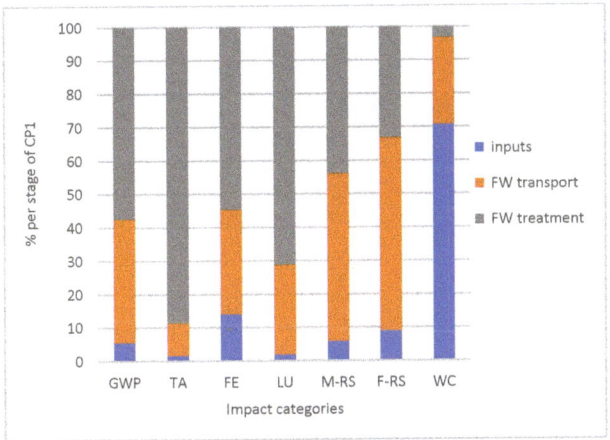

Figure 8. Impacts from FW Composting in a centralized scenario (CP1).

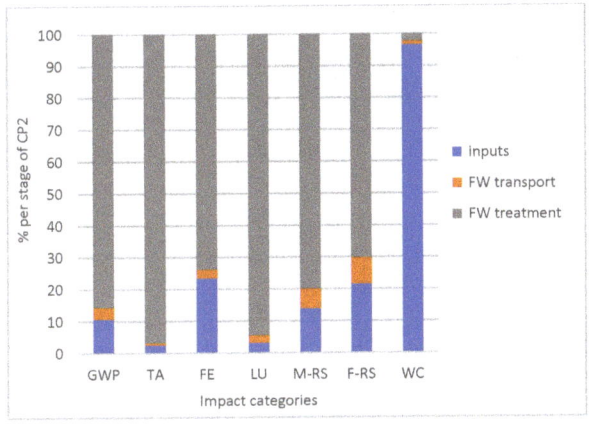

Figure 9. Impacts from FW Composting in a centralized scenario (CP2).

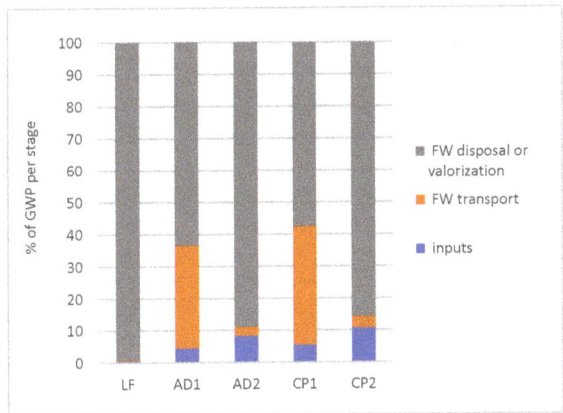

Figure 10. Contribution from each operation stage per FW treatment in the Global Warming Potential category.

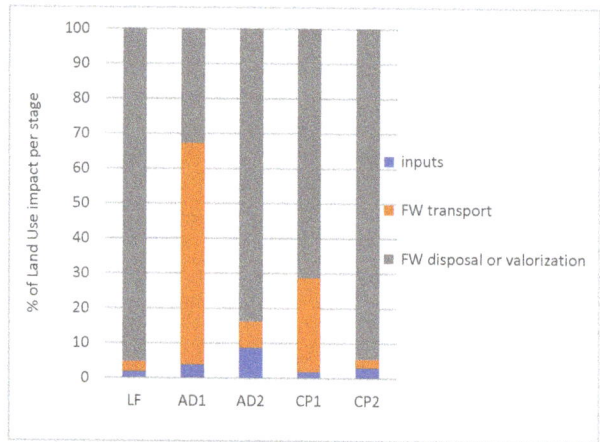

Figure 11. Contribution from each operation stage per FW treatment in the Land-Use impact category.

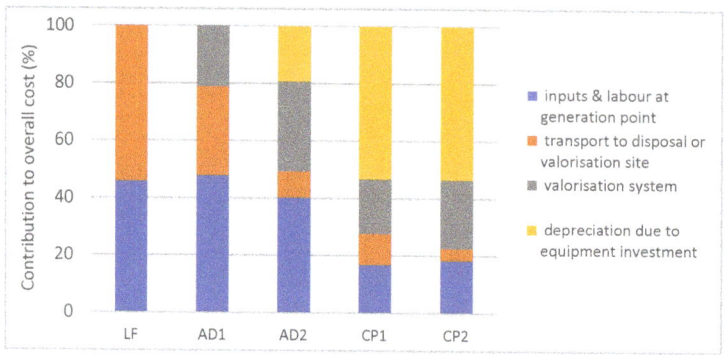

Figure 12. Proportion of each cost element within the E-LCC analysis.

Trade-offs among impact categories are observed. Transportation and the biodegradation process itself explains the higher Land-Use in CP1 and CP2; in contrast to AD1 and AD2, which would represent the least Land-Use impact of the scenarios. The Acidification Potential is influenced mostly by higher transport requirements and valorization processes. The higher potential impacts than LF for Mineral Resources and Fossil Resources depletion are attributable mostly to the increase in the FW transport. With the exception of scenario CP1, Water Consumption is also increased in all valorization alternatives, being the main reason, that now more cleaning operations would be mandatory both at generation point and in the valorization sites, while this latter was not required in the LF scenario. However, it should be observed that an approximately of 15% to 33% of this water would come from sustainable sources such as rainfall in alternative scenarios.

In summary, the FW treatment technique is the highest contributor to the Global Warming Potential, the Water Consumption and Land-Use impact categories (Figures 5–11); however, inputs would have higher proportional contributions in AD2 and CP2 scenarios, evidencing that transport becomes a hotspot.

GWP and LU are two relevant categories that would deserve in-depth observation (Figures 10 and 11), where the disposal or treatment practice plays a relevant role in the whole impact on both categories.

While transport of the FW is almost imperceptible in the overall LF Global Warming Potential (it is assumed the FW would be transported from the generation sites to closer landfills), the centralized alternatives AD1 and CP1 show a considerable increase due to the transport of the FW in route 1, accountable for a distance of 126km. This impact is lowered in the semi-centralized scenarios AD2 and CP2, where the FW transport is responsible for less than 4% of that impact, consequent with the 33km in route 2. Besides the treatment or valorization process, transport operations remain as one of the main contributors in the centralized scenarios (AD1 and CP1) for the impact category concerning Land-Use as well.

Regarding costs, the E-LCC showed that all valorization alternatives, except AD1, would result in higher yearly costs than LF (Table 5), and the contribution of the different elements of the cost will vary depending on each scenario (Figure 12).

Table 5. Cost of the business as usual and alternative FW treatment scenarios, expressed in USD.

Cost Category	LF	AD1	AD2	CP1	CP2
Inputs and labour at generation point	7635.32	7635.32	7635.32	7635.32	7635.32
Transport to disposal or valorisation plant	8986.26	4938.27	1709.29	4938.27	1709.29
Valorisation system	-	3334.81	5940.72	8562.19	9907.89
Depreciation due to equipment investment		-	3647.70	24,023.69	22,112.71
Total	16,621.58	15,908.41	18,933.03	45,159.47	41,365.21

Besides the already considered operators at the FW generation point to aid in cleaning and collection activities, AD1 and CP1 scenarios would require an estimate of 1.36 fulltime additional operators, while AD2 and CP2 would require 1.87 fulltime additional operators, with the correspondent calculation in monetary units and addition to the valorization system costs.

The increased yearly costs for most of the alternatives are attributed to a scale effect, the undertaking of new operations within the campus where the plant or plants would be established, and the depreciation of the required equipment were not previously available. LF does not incur in depreciation or valorization costs. In contrast, AD1, AD2, CP1 and CP2 would have new cost elements represented by the FW transport to the valorization site(s), and the FW processing operations, represented by materials, transport of those materials, and labor. In addition, the investment in new equipment will most definitely have economic impacts in the operation, as observed in the CP1 and CP2 scenarios in contrast to the AD1 and AD2 alternatives, or between AD1 and AD2 (for instance, AD1 would not require new investments because of an already existing and operational digestor). Similarly, CP2 would have a slightly lower depreciation cost since one of the campuses already has a composter.

The centralized scenarios such as CP1 have higher overall costs than AD2, CP2, and LF, attributable to the FW transport category; providing an important vision regarding the effects of centralization or semi-centralization of this type of recovery processes.

The study involved a system expansion, where market products were substituted by the side flows on each process, such as the biogas, digestate or compost (Table 6). For each alternative, a net effect was estimated, consisting of the impact that each new practice would suppose, the avoided impacts due to diverting the FW from the landfill, and the savings from substituting market products by the obtained side flows. In this case, liquified petroleum gas (LPG) would be substituted by biogas, and fertilizers, whether conventional or commercial compost bought by some of the universities or nearby farmers, would be substituted by the compost or digestate from the alternative scenarios.

Table 6. Net effect of FW valorization alternatives for GWP, LU and Economic effect.

Indicator		AD1	AD2	CP1	CP2
GWP (CO_2 eq)	new impact	16,113.16	11,906.42	13,973.26	9376.31
	avoided impact	1,260,078.39	1,219,486.49	91,012.29	90,050.00
	net effect	−1,243,965.24	−1,207,580.07	−7039.03	−80,673.69
LU (m^2a crop eq)	new impact	230.30	132.64	542.53	408.88
	avoided impact	13,391.95	12,248.63	417.27	388.48
	net effect	−13,161.65	−12,115.99	125.26	20.40
Net Economic effect (USD)	new cost	15,908.41	18,933.03	45,159.47	41,365.21
	avoided cost	729,659.58	729,05.66	19,985.77	20,016.95
	Net cost effect	−713,751.18	−710,772.63	25,173.70	21,348.26

The four FW-to-energy alternatives would suggest savings in CO_2 eq emissions in comparison to the business-as-usual practice. The net effect for Costs shows savings for AD1 and AD2, but avoided expenses do not make up for the potential new costs of valorizing FW in scenarios CP1 and CP2. There would be a substitution in the purchase of inputs, creating attractive yearly savings in scenarios AD1 and AD2. These savings are explained by the LPG substitution and smaller contribution from the substitution of commercial organic fertilizers once the universities use the digestate. In contrast, CP1 and CP2, even when substituting fertilizers by the obtained compost, will not represent yearly savings; instead, it will result in increased expenses because of higher depreciation costs and lower value products (compost) in regards to the AD alternatives.

A sensitivity analysis was conducted, observing effects in GWP and LU because of changes of certain inputs. On one hand, rice husk is one of the inputs for the adapted-Takakura compost method in the CP1 and CP2 alternatives, and would suppose an increased impact. However, the researchers decided to only account for its upstream transportation into the system. This was founded in the fact that the rice husk is a side flow of other processes, and the potential consumption of the input in the evaluated scenarios would not surpass 0.003% of the national inventory; therefore, competitive use of the husk or changes in the already existing local conditions are not expected to cause significant changes in the local market. Another input that deserved attention was the cleaning products, like chlorine, due to the contaminant power it is accounted for. Experts would suggest that quaternary ammonium and peracetic acid can be an effective option for disinfecting, besides the latter would be widely accepted to be used in food processing areas. Therefore, the analysis considered substituting sodium hypochlorite for acetic acids in one of the valorization scenarios, suggesting that this change would decrease the GWP by 180.976 kg CO_2 eq, and the LU by 4.224 m^2a crop eq. One additional concern in this last matter has to do with cost, since quaternary ammonium and peracetic acid can be more expensive than chlorine.

3.3. AHP Multicriteria Decision-Method

The science-to-expert approach indicated that, given the context where this case study was developed, the two most relevant criteria under which FW-to-energy alternatives should be evaluated are job generation and Land-Use (Table 7). Depreciation costs and Global Warming Potential followed at an intermediate level of relevance, and finally, the ease of implementation and the net economic effect were the less relevant criteria for these experts.

Knowing the assigned priority to the criteria, comparison matrices are presented in Table 8, consisting of the judgment for the FW-to-energy scenarios under consideration, regarding each of the evaluation criteria. Afterwards, Table 9 presents the ranking for the scenarios that entailed different FW-to-energy alternatives for this university consortium according to the experts.

Table 7. Evaluation criteria comparison matrix and priority vector.

Indicator	Global Warming Potential	Land-Use	Ease of Implementation	Job Generation	Depreciation Cost	Net Economic Effect	Priority Vector
Global Warming Potential	0.130	0.136	0.194	0.115	0.124	0.120	0.136
Land-Use	0.235	0.245	0.152	0.313	0.239	0.221	0.234
Ease of implementation	0.057	0.138	0.085	0.071	0.087	0.095	0.089
Job generation	0.306	0.212	0.327	0.271	0.301	0.297	0.286
Depreciation cost	0.178	0.173	0.165	0.152	0.169	0.181	0.170
Net economic effect	0.093	0.095	0.077	0.078	0.080	0.086	0.085

CR = 0.01 < 0.10

Table 8. FW-to-energy valorization alternatives comparison matrices for each evaluation criterion.

	Global Warming Potential				Land-Use			
	AD1	AD2	CP1	CP2	priority vector	AD1	AD2	CP1	CP2	priority vector
AD1	0.201	0.278	0.204	0.163	0.211	0.266	0.391	0.255	0.176	0.272
AD2	0.107	0.149	0.166	0.169	0.148	0.108	0.159	0.285	0.110	0.166
CP1	0.405	0.368	0.411	0.435	0.405	0.360	0.194	0.346	0.538	0.360
CP2	0.287	0.205	0.220	0.233	0.236	0.266	0.256	0.113	0.176	0.203
					CR = 0.02					CR = 0.09

	Ease of Implementation					Job Generation				
	AD1	AD2	CP1	CP2	priority vector	AD1	AD2	CP1	CP2	priority vector
AD1	0.257	0.265	0.335	0.178	0.259	0.542	0.540	0.576	0.484	0.535
AD2	0.251	0.258	0.290	0.210	0.252	0.187	0.186	0.179	0.189	0.185
CP1	0.190	0.220	0.247	0.402	0.265	0.163	0.180	0.173	0.230	0.186
CP2	0.302	0.258	0.129	0.210	0.225	0.108	0.094	0.072	0.096	0.093
					CR = 0.05					CR = 0.01

	Depreciation Cost					Net Economic Effect				
	AD1	AD2	CP1	CP2	priority vector	AD1	AD2	CP1	CP2	priority vector
AD1	0.199	0.110	0.268	0.208	0.196	0.221	0.385	0.191	0.186	0.246
AD2	0.331	0.182	0.233	0.095	0.210	0.093	0.161	0.254	0.156	0.166
CP1	0.246	0.260	0.332	0.464	0.325	0.443	0.242	0.382	0.453	0.380
CP2	0.224	0.447	0.167	0.234	0.268	0.243	0.211	0.173	0.205	0.208
					CR = 0.10					CR = 0.06

Table 9. Evaluation and ranking of the FW-to-energy alternatives under study.

(priority vector)	Global Warming Potential (0.136)	Land-Use (0.234)	Ease of Implementation (0.089)	Job Generation (0.286)	Depreciation Cost (0.170)	Net Economic Effect (0.085)	Prioritization	Ranking
AD1	0.029	0.064	0.023	0.153	0.033	0.021	0.323	1
AD2	0.020	0.039	0.022	0.053	0.036	0.014	0.184	4
CP1	0.055	0.084	0.024	0.053	0.055	0.032	0.304	2
CP2	0.032	0.048	0.020	0.026	0.046	0.018	0.189	3

4. Discussion

As a first method of the proposed methodological framework, the calculations through the site decision matrix and linear programming allowed the researchers to identify site E as the preferred to establish an FW valorization plant. However, it is observed that factors such as the integration of the consortium and the withdrawing of one of them from the route would account for a significant reduction of distance between one route and the other. Expected implications were observed in the results of CP or AD alternatives, entailing effects on the environmental and cost performance of the valorization alternatives, as well as in the decision in this study, of centralized FW valorization systems (with an FW collection route of 126 km) or semi-centralized ones (with FW collection route of 33km). In this sense, careful selection of parameters, weighting, and available information play a key role in the output of similar design of case studies.

As the second method of the framework, LCT proves to be clear and consistent in expressing the environmental and cost impacts of the evaluated scenarios. Previous sources indicate the fittingness of LCA [27,42] and E-LCC [15,25,42] to approach waste management situations, including FW. The stability of the LCT for the environmental and economic dimensions of sustainability studies, oriented by ISO14040 Standard [48] and Hunkeler D., Lichtenvort K., and Rebitzer, G [49] opens the possibility of comparability, perhaps not always among cases due to the diversity of elements of each scenario, but within cases as an improvement monitoring tool. However, the social dimension is still not addressed in the same manner, suggesting this to be a further area of research. Therefore, in this case, it was mostly evaluated by experts.

The LCA and E-LCC results of this study suggest AD to be the better performing FW-to-energy alternative, whether centralized or semi-centralized, being consistent with the municipal and experimental analysis that locates anaerobic digestion as a suitable treatment in terms of lower environmental impacts, side flow opportunities and economic perspectives [14,15,34]. Even when finding coincidence in hotspots such as the actual degradation technique, many of the consulted sources disregard transportation as a hotspot, reinforcing the need to observe system boundaries definitions when comparing studies and the relevance of centralization (or not) when proposing waste management systems [17].

The properties of the biomass to be valorized play a relevant role in the outcomes of each alternative. This study undertook valorization techniques already proven to work under the local conditions [21,22], based particularly in the balance of food groups comprised in FW. However, further experimentation based on properties regarding side flow production, calorific or nutritional potential and feedstock [17], as well as geographical origin, the type of collection source and the season of the collection [20] should be considered. In this sense, biomass characteristics could be included inside flow characterization experiments, or as a criterion to be assessed by the experts in decisional methods like the AHP.

The E-LCC also allowed to observe that the contribution of each element of the cost will vary with each scenario and more long-term and expanded perspectives must become part of these decisions. As an illustration of this argument, it is possible that if the decision was to rest uniquely upon the overall cost, FW treatment alternatives would not be of interest due to higher costs than the business-as-usual scenario. However, a wider comprehension of the circular economy principles, that considers the use and value of side flows [2,42] and performed by a system expansion in this study, allowed to understand that the AD scenarios would not only be avoiding environmental impacts but would be generating potential incomes.

Context-wise, it is relevant to highlight that the biogas was not considered in this case for electricity production, since the Costa Rican electricity grid is considered as already sustainable, and sufficient energy comes mostly from hydroelectric sources, followed by Eolic and geothermal sources [58]. Nonetheless, the country has an important consumption of fossil fuels such as LPG for combustion. For instance, universities would use liquefied petroleum gas (LPG) for their academic and research laboratories, and in their restaurants. In parallel, there is a relevant amount of Costa Ricans that would use LPG for cooking [56], consequently the study assumed the universities or nearby users can substitute LPG by biogas; however, acceptance rates were not inserted in this study. Another product would be the digestate from the AD1 and AD2 scenarios, as well as the compost in CP1 and CP2 scenarios. The selected valorization sites could use these products as a source of organic fertilizers in experimental fields where agricultural-related study programs or gardening activities are detected. Consequently, the costs and substitution preference cannot be considered as general for these valorization alternatives, but rather case-specific.

As a third and final step in the proposed decisional framework, the AHP method allowed the consulted experts to provide answers that were later computed to rank the alternatives. In this case, the two centralized alternatives ranked first (AD1 followed by CP1), then the semi-centralized composting scenario CP2 obtained the third place in the ranking and the AD2 alternative was placed fourth. Both the results of the pairwise comparisons and the open-question answers suggest AD1 would have a priority within the alternatives because of less Land-Use impact and lower Depreciation costs. However, when considering the rest of the criteria, CP1 was a second choice related to aspects such as the Ease-of-implementation. This last criterion, even when not highly prioritized, was usually present in the comments of the experts, mentioning that operating one facility might be easier than managing the simultaneous operation of several. In that sense, the comparison between the two semi-centralized alternatives, suggested composting in three plants was preferred to installing AD plants in three sites, therefore AD2 ranked the lowest from the four alternatives.

Feedback from the experts resulted in a positive overview of the proposed methodological framework for decision-making towards more circular approaches to manage FW, given the combination of methods and quantitative data that allowed to better understand the scenarios when supplied to the experts. They also expressed the sequence of methods allowed them to make an informed choice together with their experience and knowledge. Finally, they also found it to be innovative for the local context where decisions need to be more robust and consistent, since public policy creation, and implementation is considered by them to be a complex, multidisciplinary and dynamic process. Other experts suggested future scenarios to be evaluated as well, due to the scale of the consortium, where more artisanal composters were evaluated, and some presented a potential concern regarding the use of biogas and its acceptability at the consortium and local levels.

5. Conclusions

This paper evaluated four FW-to-energy alternatives and compared them to a landfill scenario through a system expanded LCA and E-LCC. The ultimate purpose was to contribute in decision-making processes related to FW valorization alternatives, and therefore it proposed an integrated methodological framework, combining LCT approaches with Linear programming and multicriteria decision methods such as (AHP).

From the environmental standpoint, main findings indicate that FW valorization alternatives in general, would entail reduced Global Warming Potential and Freshwater Eutrophication than the landfilling alternative; however, trade-offs are observed regarding other impact categories such as Terrestrial Acidification, Mineral Resource Scarcity and Fossil Resource Scarcity, where the potential impact from the valorization would be increased. Other environmental impact categories would perform differently when anaerobic digestion or composting were evaluated; nonetheless, it was clear that anaerobic digestion would entail lower Land-Use than composting and landfilling. Moreover, centralization or semi-centralization would also suggest different impacts, mostly in terms of the contribution that transportation would make to each impact category.

Regarding the economic and social dimensions, the findings conclude that, for the given circumstances and context, most of the FW-to-energy alternatives would have higher overall costs than the landfilling, something that is evidently reverted once a system expansion approach is considered. In this sense, when the valorization and the circular economy concepts are understood and explained through savings in products that can be substituted by side flows of the composting and the anaerobic digestion of the wastes, the proposed alternatives can become appealing for decision-makers. Besides, the valorization of the FW would require more labor, seen as an increased cost but also as an opportunity for job creation.

Further research and validation of the framework in different contexts are suggested, as well as the consideration of extended scopes where other criteria are evaluated, such as more in-depth biomass composition and energy properties, and the effects on the obtained side flows.

The trade-offs and potential interpretation of results will not always provide a straightforward selection of an alternative. Therefore, the proposed holistic methodological framework allowed a logical process of case definition and scenarios modeling, accompanied by scientifically-based assessment methods, together with a science-to-expert approach. This latter comprised a better understanding within this context, once experts offer their perspective by a well-structured and systematized method as the AHP.

Even with the limits of a case study, this research suggests that the circular economy is applicable for different activities. Evidence is always necessary to consider shifting from one scenario, as the usual and current landfilling one, to a more circular one where valorization of FW can improve not only the waste management within this university consortium, but the obtention of valuable products with the opportunity to positively affect environmental, social, and economic indicators.

Similar cases, such as small municipalities or groups of institutions, can benefit from a similar approach as the one presented in this research, since decisions can be guided in a systematic manner with already proven, sequential and steady methods like linear programming, LCA, LCC and AHP.

Supplementary Materials: The following are available online at http://www.mdpi.com/1996-1073/13/9/2291/s1 and include LCA and LCC information (Goal and Scope, system boundaries and flow diagram, inventories of LF, CP1, CP2, AD1 and AD2 scenarios, calculations and assumptions for inventories and generic information and sources used for those calculations.

Author Contributions: Conceptualization: L.B.-P.; methodology: L.B.-P., F.D.M.; M.F.J.-M.; validation, M.V. and R.C.-R.; formal analysis, L.B.-P., F.D.M. and M.F.J.-M.; writing—original draft preparation L.B.-P.; writing—review and editing, M.V., F.D.M. and R.C.-R.; supervision, M.V. All authors have read and agreed to the published version of the manuscript.

Funding: This research was funded by Tecnológico de Costa Rica, project ID 1431012 in cooperation with the Universitly of Bologna within the Agricultural, Environmental and Food Science and Technology PhD Programme, CV in International cooperation and sustainable development policies.

Acknowledgments: The authors would like to thank and recognized the support provided by Gerlin Salazar from University of Costa Rica, Oliviero Bergamin from the University of Bologna, and Rui Leonardo Madime, Felipe Vaquerano, Rubén Calderón, from Tecnológico de Costa Rica for their professional advice. Students Jonathan Castro, Daniela Valverde, Marianela Ávila, Raizeth Chaves, Noelia González, Fiorella Ramírez, Eva Vargas, Andrey Ureña and Rolando Jimenez from Tencnológico de Costa Rica also provided provided support during field and data collection processes. We would also like to recognize REDIES for their proactive approach in FW quantification together with the Costa Rican Food Loss and Waste Network, together with the authors of the first FW quantification study in Costa Rica: Julián Rojas-Vargas, Yanory Monge-Fernán, Manrique Arguedas Camacho, Cindy Hidalgo-Viquez, Marcela Peña-Vásquez y, Blanca Vásquez Rodríguez, co-authors with Laura Brenes-Peralta and María Fernanda Jiménez-Morales. Finally, the experts of the consultation provided invaluable support to this research, out deep thanks.

Conflicts of Interest: The authors declare no conflict of interest.

References

1. Corona, B.; Shen, L.; Reike, D.; Carreón, J.R.; Worrell, E. Towards sustainable development through the circular economy—A review and critical assessment on current circularity metrics. *Resour. Conserv. Recycl.* **2019**, *151*, 104498. [CrossRef]
2. Ellen MacArthur Foundation. Concept, What Is Circular Economy? A Framework for an Economy that Is Restorative and Regenerative by Design. Available online: https://www.ellenmacarthurfoundation.org/circular-economy/concept (accessed on 18 December 2019).

3. Fridahl, M. Bioenergy with Carbon Capture and Storage: From Global Potentials to Domestic Realities. Available online: https://www.liberalforum.eu/wp-content/uploads/2018/11/beccs_publication.pdf (accessed on 5 October 2019).
4. Herrera-Murillo, J.; Rojas-Marín, J.F.; Anchía-Leitón, D. Tasas de generación y caracterización de residuos sólidos ordinarios en cuatro municipios del área metropolitana costa rica. *Revista Geográfica de América Central* **2016**, *2*, 235–260. [CrossRef]
5. FAO. Food Wastage Footprint & Climate CHANGE. Available online: http://www.fao.org/3/a-bb144e.pdf (accessed on 5 October 2019).
6. FAO. *Food Wastage Footprint: Fool Cost-Accounting, Final Report*; FAO: Rome, Italy, 2014; ISBN 978-92-5-108512-7.
7. Corrado, S.; Ardente, F.; Sala, S.; Saouter, E. Modelling of food loss within life cycle assessment: From current practice towards a systematisation. *J. Clean. Prod.* **2017**, *140*, 847–859. [CrossRef]
8. FAO. *FAO and the SDSs. Indicators: Measuring up to the 2030 Agenda for Sustainable Development*; FAO: Rome, Italy, 2017.
9. HLPE. *Food Losses and Waste in the Context of Sustainable Food Systems. A Report by the High Level Panel of Experts on Food Security and Nutrition of the Committee on World Food Security*; FAO: Rome, Italy, 2014.
10. Wang, D.; Tang, Y.-T.; Long, G.; Higgitt, D.; He, J.; Robinson, D. Future improvements on performance of an EU landfill directive driven municipal solid waste management for a city in England. *Waste Manag.* **2020**, *102*, 452–463. [CrossRef]
11. Papargyropoulou, E.; Lozano, R.; Steinberger, J.K.; Wright, N.; Bin Ujang, Z. The food waste hierarchy as a framework for the management of food surplus and food waste. *J. Clean. Prod.* **2014**, *76*, 106–115. [CrossRef]
12. Hoehn, D.; Margallo, M.; Laso, J.; Aldaco, R.; Bala, A.; Fullana-I-Palmer, P.; Irabien, A.; Aldaco, R. Energy Embedded in Food Loss Management and in the Production of Uneaten Food: Seeking a Sustainable Pathway. *Energies* **2019**, *12*, 767. [CrossRef]
13. El Mashad, H.M.; Zhang, R. Biogas production from co-digestion of dairy manure and food waste. *Bioresour. Technol.* **2010**, *101*, 4021–4028. [CrossRef]
14. Slorach, P.C.; Jeswani, H.K.; Cuéllar-Franca, R.; Azapagic, A. Environmental and economic implications of recovering resources from food waste in a circular economy. *Sci. Total Environ.* **2019**, *693*. [CrossRef]
15. Edwards, J.; Othman, M.; Crossin, E.; Burn, S. Life cycle assessment to compare the environmental impact of seven contemporary food waste management systems. *Bioresour. Technol.* **2018**, *248*, 156–173. [CrossRef]
16. Demichelis, F.; Piovano, F.; Fiore, S. Biowaste Management in Italy: Challenges and Perspectives. *Sustainability* **2019**, *11*, 4213. [CrossRef]
17. Xu, F.; Li, Y.; Ge, X.; Yang, L.; Li, Y. Anaerobic digestion of food waste—Challenges and opportunities. *Bioresour. Technol.* **2018**, *247*, 1047–1058. [CrossRef] [PubMed]
18. Ávila-Hernández, M.; Campos-Rodríguez, R.; Brenes-Peralta, L.; Jiménes-Morales, M.F. Generación de biogás a partir del aprovechamiento de residuos sólidos biodegradables en el Tecnológico de Costa Rica, sede Cartago. *Revista Tecnología en Marcha* **2018**, *31*, 159–170. [CrossRef]
19. Bong, C.P.C.; Lim, L.Y.; Lee, C.T.; Klemeš, J.J.; Ho, C.S.; Ho, W.S. The characterisation and treatment of food waste for improvement of biogas production during anaerobic digestion—A review. *J. Clean. Prod.* **2018**, *172*, 1545–1558. [CrossRef]
20. Fisgativa, H.; Tremier, A.; Dabert, P. Characterizing the variability of food waste quality: A need for efficient valorisation through anaerobic digestion. *Waste Manag.* **2016**, *50*, 264–274. [CrossRef] [PubMed]
21. Borrero-González, G.P.; Arias-Aguilar, D.; Campos-Rodríguez, R.; Pacheco-Rodríguez, F. Estudio comparativo del uso de dos sustratos con inóculos microbiales para el tratamiento de residuos orgánicos sólidos en compostaje doméstico. Análisis Económico. *Revista Tecnología en Marcha* **2016**, *29*, 28–37. [CrossRef]
22. Chaves-Arias, R.; Campos-Rodríguez, R.; Brenes-Peralta, L.; Jiménez-Morales, M.F. Compostaje de residuos sólidos biodegradables del restaurante institucional del Tecnológico de Costa Rica. *Revista Tecnología en Marcha* **2019**, *232*, 39–53. [CrossRef]
23. Ramírez-Ramírez, F.; Campos-Rodríguez, R.; Jiménez-Morales, M.F.; Brenes-Peralta, L. Evaluación técnica, ambiental y económica de tres tipos de tratamiento para el cultivo de lechuga en huertas caseras de Guácimo, Limón, Costa Rica. *Revista Tecnología en Marcha* **2016**, *29*, 14–24. [CrossRef]
24. Yang, Y.; Bao, W.; Xie, G.H. Estimate of restaurant food waste and its biogas production potential in China. *J. Clean. Prod.* **2019**, *211*, 309–320. [CrossRef]

25. De Menna, F.; Dietershagen, J.; Loubière, M.; Vittuari, M. Life cycle costing of food waste: A review of methodological approaches. *Waste Manag.* **2018**, *73*, 1–13. [CrossRef]
26. Cleary, J. The incorporation of waste prevention activities into life cycle assessments of municipal solid waste management systems: Methodological issues. *Int. J. Life Cycle Assess.* **2010**, *15*, 579–589. [CrossRef]
27. Saraiva, A.B.; Jansen, J.L.C. Review of comparative LCAs of food waste management systems—Current status and potential improvements. *Waste Manag.* **2012**, *32*, 2439–2455. [CrossRef]
28. Saaty, T.L. Decision making with the analytic hierarchy process. *Int. J. Serv. Sci.* **2008**, *1*, 83–98. [CrossRef]
29. Campos-Rodríguez, R.; Soto-Córdoba, S. Análisis de la situación del estado de la Gestión Integral de Residuos (GIR) en el Cantón de Guácimo, Costa Rica. *Revista Tecnología en Marcha* **2014**, *27*, 114–124. [CrossRef]
30. Rojas-Vargas, J.; Monge-Fernánde, Y.; Jiménez-Morales, M.F.; Brenes-Peralta, L.; Arguedas-Camacho, M.; Hidalgo-Víquez, C.; Peña-Vásquez, M.; Vasquez-Rodríguez, B. Food Loss and Waste in food services from educational institutions in Costa Rica. *Tecnología en Marcha* **2020**. (accepted).
31. Sepúlveda, J.A.M. Outlook of Municipal Solid Waste in Bogota (Colombia). *Am. J. Eng. Appl. Sci.* **2016**, *9*, 477–483. [CrossRef]
32. JICA; Ex Reserach Institute Ltd.; Rurban Designs Inc. Estudio y Recopilación de Datos sobre el Sector de Manejo de Residuos Sólidos en América Central y Caribe. Available online: https://openjicareport.jica.go.jp/pdf/12091906.pdf (accessed on 30 March 2020).
33. Sánchez Tejeda, G.M. Análisis de Ciclo de Vida aplicado a la gestión de residuos urbanos del Distrito Nacional de la República Dominicana. Master's Thesis, Nebrija Universidad, Madrid, Spain, July 2012.
34. Fernández-González, J.; Grindlay, A.; Serrano-Bernardo, F.; Rojas, M.I.R.; Zamorano, M. Economic and environmental review of Waste-to-Energy systems for municipal solid waste management in medium and small municipalities. *Waste Manag.* **2017**, *67*, 360–374. [CrossRef] [PubMed]
35. Rodríguez, A.G.; Rodrigues, M.; Sotomayor, O. *Hacia una Bioeconomía Sostenible en América Latina y el Caribe: Elementos para una Visión Regional. Serie Recursos Naturales y Desarrollo*; Comisión Económica para América Latina y el Caribe (CEPAL): Santiago, Chile, 2019.
36. FAO. *4° Boletín Pérdidas y Desperdicio de Alimentos en América Latina y el Caribe*; FAO-RLC: Santiago, Chile, 2017; I7248ES/1/12.17.
37. MINAE; MIDEPLAN; Ministerio de Relaciones Exteriores de Costa Rica. Política Nacional de Producción y Consumo Sostenibles 2018–2030. Available online: http://www.digeca.go.cr/sites/default/files/documentos/politica_nacional_de_produccion_y_consumo_sostenibles.pdf (accessed on 8 January 2020).
38. MINAE. Plan de Descarbonización 2019–2050. Available online: https://minae.go.cr/images/pdf/Plan-de-Descarbonizacion-1.pdf (accessed on 8 January 2020).
39. Sistema Costarricense de Información Jurídica, Ley para la Gestión Integral de Residuos 883. Available online: https://www.pgrweb.go.cr/scij/Busqueda/Normativa/Normas/nrm_texto_completo.aspx?param1=NRTC&nValor1=1&nValor2=68300&nValor3=83024&strTipM=TC (accessed on 8 January 2020).
40. Red Costarricense para Disminución de pérdida y desperdicio de alimentos. Guía para Medición de Desperdicio de Alimentos en Cocinas Institucionales o Comerciales. Available online: https://www.tec.ac.cr/sites/default/files/media/doc/2_guia_medicion_cocinas_web.pdf (accessed on 20 January 2020).
41. Östergren, K.; Gustavsson, J.; Bos-Brouwers, H.; Timmermans, T.; Hansen, O.-J.; Møller, H.; Anderson, G.; O'Connor, C.; Soethoudt, H.; Quested, T.; et al. *FUSIONS Definitional Framework for Food Waste*; SIK—The Swedish Institute for Food and Biotechnology: Göteborg, Sweden, 2014; ISBN 978-91-7290-331-9.
42. Davis, J.; De Menna, F.; Unger, N.; Östergren, K.; Loubiere, M.; Vittuari, M. *Generic Strategy LCA and LCC —Guidance for LCA and LCC Focused on Prevention, Valorisation and Treatment of Side Flows from the Food Supply Chain*; SP Technical Research Institute of Sweden: Boras, Sweden, 2017; ISBN 978-91-88349-84-2.
43. Tobón Botache, M.I.; Cruz Viveros, J.A. Métodos de Localización de Plantas Industriales. Available online: https://repository.usc.edu.co/handle/20.500.12421/2458 (accessed on 14 February 2020).
44. Carro Paz, R.; González Gómez, D. Localización de Instalaciones. Universidad Nacional de Mar del Plata. Available online: http://nulan.mdp.edu.ar/1619/1/14_localizacion_instalaciones.pdf (accessed on 14 February 2020).
45. Quintero-Peralta, M.A.; Gallardo-Cobos, R.; Ceña-Delgado, F. Impact of decreasing staple food production capacity on food self-sufficiency in poor rural communities in Mexico. *Economía Agraria y Recursos Naturales* **2016**, *16*, 33–67. [CrossRef]

46. Gaytán Iniesta, J.; Arroyo López, M.d.P.E.; Enríquez Colón, R. Un modelo bi-criterio para la ubicación de albergues, como parte de un plan de evacuación en caso de inundaciones. *Revista Ingeniería Ind.* **2012**, *11*, 35–56.
47. Taha, H.A. *Investigación de Operaciones, Authorized Translation from the English Language Edition Entitled Operations Research: An Introduction*, 9th ed.; Pearson Education Inc.: Ciudad de México, Mexico, 2012; ISBN 978-607-32-0796-6.
48. INTECO. *Análisis de ciclo de vida: INTE/ISO 14040:2007*; INTECO: San José, Costa Rica, 2007.
49. Hunkeler, D.; Lichtenvort, K.; Rebitz, G. *Environmental Life Cycle Costing*; Taylor & Francis Group: Boca Raton, FL, USA, 2008; ISBN 1-880611-38-X.
50. De Menna, F.; Davis, J.; Östergren, K.; Unger, N.; Loubiere, M.; Vittuari, M. A combined framework for the life cycle assessment and costing of food waste prevention and valorization: An application to school canteens. *Agric. Food Econ.* **2020**, *8*, 1–11. [CrossRef]
51. Brenes-Peralta, L.; Jiménez-Morales, M.F. Condición actual del mercado del abono orgánico en el cantón de Alvarado, Cartago. *Revista Tecnología en Marcha* **2014**, *27*, 65–75. [CrossRef]
52. Ministerio de Salud de Costa Rica. Inventario de Georeferenciación y de Caracterización Fisico—Químico de Lixiviados, Suelos y Gases, en sitios de Disposicion Final de Residuos. Available online: https://www.ministeriodesalud.go.cr/index.php/component/content/article?id=617 (accessed on 8 January 2020).
53. Rudín, V.; Soto, S.; Linnenberg, C. Elaboración de la Propuesta de Proyecto a Financiar para una NAMA de Residuos Sólidos en Costa Rica. Available online: https://cambioclimatico.go.cr/wp-content/uploads/2019/07/Primer-informe-Situaci%C3%B3n-de-la-Gesti%C3%B3n-de-los-Residuos-S%C3%B3lidos-para-la-determinaci%C3%B3n-de-la-NAMA-residuos-Costa-Rica.pdf (accessed on 20 January 2020).
54. Suhartini, S.; Lestari, Y.P.; Nurika, I. Estimation of methane and electricity potential from canteen food waste. *Int. Conf. Green Agro Ind. Bioecono. IOP Conf. Ser. Earth Environ. Sci.* **2019**, *230*. [CrossRef]
55. United Nations. Conversion Factors. Available online: http://mdgs.un.org/unsd/energy/balance/2013/05.pdf (accessed on 8 January 2020).
56. RECOPE. Gas Licuado de Petróleo (G.L.P.). Available online: https://www.recope.go.cr/productos/calidad-y-seguridad-de-productos/gas-licuado-de-petroleo-glp/ (accessed on 8 January 2020).
57. Bergamin, O. Evaluation of the Economic and Environmental Feasibility of a Biogas Plant in a University Campus: The Case of Tecnológico: The Case of Tecnológico de Costa Rica. Master's Thesis, University of Bologna, Italy, February 2019.
58. ICE. Informe Anual 2018 Generación y Demanda, Centro Nacional de Control de Energía. Available online: https://apps.grupoice.com/CenceWeb/CenceDescargaArchivos.jsf?init=true&categoria=3&codigoTipoArchivo=3008 (accessed on 30 March 2020).

© 2020 by the authors. Licensee MDPI, Basel, Switzerland. This article is an open access article distributed under the terms and conditions of the Creative Commons Attribution (CC BY) license (http://creativecommons.org/licenses/by/4.0/).

MDPI
St. Alban-Anlage 66
4052 Basel
Switzerland
Tel. +41 61 683 77 34
Fax +41 61 302 89 18
www.mdpi.com

Energies Editorial Office
E-mail: energies@mdpi.com
www.mdpi.com/journal/energies

www.ingramcontent.com/pod-product-compliance
Lightning Source LLC
LaVergne TN
LVHW070627100526
838202LV00012B/749